Data-Driven Fault Diagnosis

Data-Driven Fault Diagnosis: A Machine Learning Approach for Industrial Components delves into the application of machine learning techniques for achieving robust and efficient fault diagnosis in industrial components.

The book covers a range of key topics, including data acquisition and preprocessing, feature engineering, model selection and training, and real-time implementation of diagnostic systems. It examines popular machine learning algorithms such as support vector machines, convolutional neural networks, and extreme learning machines, highlighting their strengths and limitations in different industrial contexts. Practical case studies and real-world examples from various sectors illustrate the real-world impact of these techniques.

The aim of this book is to empower engineers, data scientists, and researchers with the knowledge and tools necessary to implement data-driven fault diagnosis systems in their respective industrial domains.

Govind Vashishtha received a PhD degree in Mechanical Engineering from the Sant Longowal Institute of Engineering and Technology, Longowal, India, in 2022. He is currently working as a Visiting Professor at Wroclaw University of Science and Technology, Wroclaw, Poland. He has authored over 70 research papers in Science Citation Index (SCI) journals and has also edited one book. His name has appeared in the world's top 2% scientist list published by Stanford University in 2023 and 2024. He is also serving as Associate Editor in *Frontiers in Mechanical Engineering, Shock and Vibration, Measurement and Engineering*, and *Applications of Artificial Intelligence*. He has two Indian patents. His H-index is 27 and he has more than 1800 citations to his credit. His current research includes fault diagnosis of mechanical components, vibration and acoustic signal processing, identification/measurement, defect prognosis, machine learning, and artificial intelligence.

Data-Driven Fault Diagnosis

A Machine Learning Approach for Industrial Components

Govind Vashishtha

CRC Press
Taylor & Francis Group
Boca Raton London New York

CRC Press is an imprint of the
Taylor & Francis Group, an **Informa** business

First edition published 2026
by CRC Press
2385 NW Executive Center Drive, Suite 320, Boca Raton FL 33431

and by CRC Press
4 Park Square, Milton Park, Abingdon, Oxon, OX14 4RN

CRC Press is an imprint of Taylor & Francis Group, LLC

© 2026 Govind Vashishtha

ISBN: 978-1-041-01163-7 (hbk)
ISBN: 978-1-041-01455-3 (pbk)
ISBN: 978-1-003-61482-1 (ebk)

DOI: 10.1201/9781003614821

Typeset in Times
by SPi Technologies India Pvt Ltd (Straive)

Contents

1 Introduction

1.1 CONDITION MONITORING (CM)

The condition monitoring (CM) strategy can detect various defects affecting the performance of hydraulic machines. CM aims to provide early warnings of potential equipment failures to facilitate timely diagnosis and repair [1–3]. Monitoring various machine parameters (vibration, acoustics, temperature, oil condition, electrical parameters, corrosion, etc.) helps assess its overall health [4]. Additionally, models can be established using different condition-monitoring techniques to simulate and predict the change in the behavior of parameters [5]. Prior knowledge of these measurement parameters helps the maintenance engineer to improve the performance of the machine. CM offers the following advantages:

1. Unexpected catastrophic breakdowns can be avoided, which may have expensive or dangerous consequences.
2. Production time/available time to a machine is increased which ultimately cuts the maintenance cost.
3. Unnecessary intervention in the functioning of a healthy machine can be eliminated.
4. It reduced the consumption of extra power by correcting the fault in time.

In condition-based maintenance, the data gathered from the monitored machine is collected, processed, and analyzed to assess its health condition. Based on the analysis done, a replacement or repair decision is taken. Improvements in sensors, data collection, signal processing, and appropriate software make this approach more effective. That's why the CM system is capable enough to predict accurate and precise information about the machine's health condition even in the presence of environmental noise, electrical interference, severe deterioration, faults, etc. The type of defect, its severity, and the location of such defect within the machine can easily be identified. The CM system can also predict the approximate remaining useful life of the machine. The CM techniques are categorized in the following ways:

1.1.1 VIBRATION-BASED CONDITION MONITORING

Vibration-based CM employs noninvasive sensors and data capture to evaluate machine performance in both time and frequency domains. Variations in these attributes indicate possible damage or deterioration [6].

Defects can be of different types such as misalignment, unbalance, mechanical looseness, spall, and pitting. Analysis of the signal warns the maintenance department about possible failures and measures to control the operating process parameters of the system. The vibration-based CM uses the concept that the damage in the system alters the system's mass, stiffness, and dissipation properties, resulting in a change in the system's dynamic response. However, it has a limitation that sometimes local damages may not affect the low-frequency responses making diagnosis difficult [7].

1.1.2 Visual Inspection

Visual inspection, using human senses (sight, hearing, touch, smell) or simple tools, provides a flexible and readily accessible way to assess a system's condition [8, 9].

1.1.3 Temperature Monitoring

Defects in machine components result in an increase in friction which produces heat. In temperature monitoring, tracking the temperature of the lubricant or housing of different components is considered.

1.1.4 Acoustic-Based Condition Monitoring

Acoustic-based CM is a form of nondestructive testing that examines the acoustic or noise waveforms produced by machinery. The sound produced by machinery components is acquired by the microphone to predict its health condition. Generally, microphones can be easily installed compared to sensors and have a high-frequency response range [10, 11].

1.1.5 Acoustic Emission in Condition Monitoring

In acoustic emissions, a strain energy release occurs rapidly, creating an elastic wave when there is deformation or damage on the surface of the machinery component. There are different acoustic emission sources in rotating machinery such as friction, turbulence, defects, cavitation, and fatigue. The most common parameters which help in acoustic emission CM are root mean square, energy, and Kurtosis [12].

1.1.6 Oil Based Condition Monitoring

In oil-based CM, rough working conditions lead to changes in physicochemical properties, which help determine the health status of the machinery. The slow degradation process of rotating machinery can also be monitored by oil analysis. In oil-based CM, the actual degree of degradation is difficult to evaluate [13].

1.2 IMPORTANCE OF OPERATING FREQUENCY IN DEFECT IDENTIFICATION

Hydraulic flow-induced rotary systems are prone to different defects such as a broken impeller or bucket, clogging, added mass, cavitation, misalignment in the shaft, and bearing defects. Such defects show its characteristic signature related to shaft speed [14, 15]. For instance, the imbalance in the hydraulic system such as a Pelton wheel and centrifugal pump causes the rotor to vibrate at shaft frequency [16]. Similarly, misalignment defects raise the frequency sometimes equal to shaft frequency and sometimes more than that depending on the type of misalignment [17, 18].

1.3 HYDRAULIC FLOW-INDUCED ROTARY SYSTEM

A fluid energy system generally converts one form of energy into another. On the basis of the direction of conversion of energy, fluid machines are further classified. Devices that transform stored energy (such as kinetic, potential, and intermolecular energy) into mechanical work are referred to as turbines. In contrast, devices that use the mechanical energy from moving components to enhance the stored energy of a fluid are called pumps, blowers, and fans. Pelton turbine is generally used in utilizing high-head energy applications. The buckets are attached to the rotor that is in the form of a circular disk and are driven by the perpendicular jet delivered through one or more jets (Figure 1.1).

Francis turbines are reaction turbines, unlike Pelton turbines. A key difference is that pressure drop in a Francis turbine occurs both before and within the runner, due to its diverging flow path, whereas in a Pelton turbine, the entire pressure drop happens within the runner. Also, a Pelton turbine's buckets interact with the water jet individually, while a Francis turbine's runner is fully submerged (Figure 1.2).

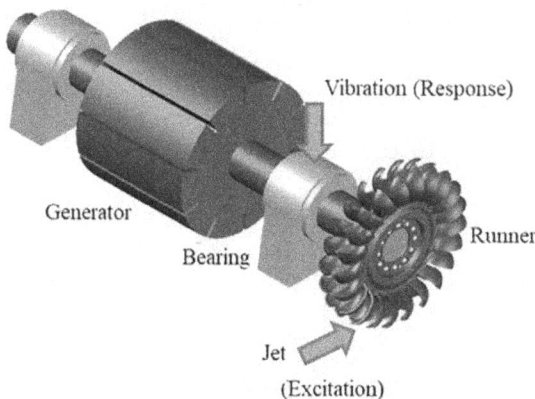

FIGURE 1.1 Pelton turbine.

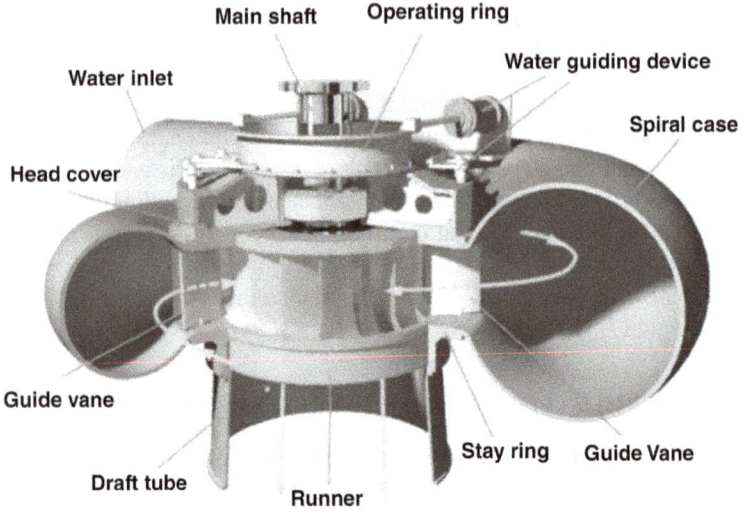

FIGURE 1.2   Francis turbine.

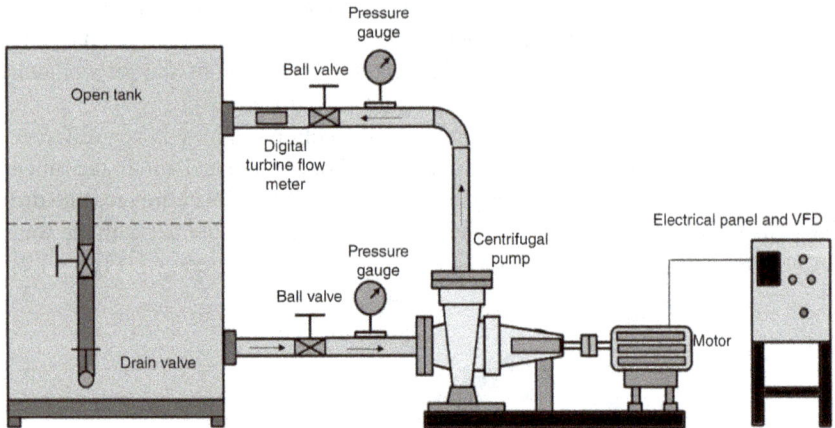

FIGURE 1.3   Centrifugal pump.

Centrifugal pumps convert rotational energy (from a motor or engine) into fluid flow energy. As a type of turbomachinery, they accelerate fluid entering near their axis outward through an impeller into a diffuser or volute casing, increasing the fluid's pressure and velocity (Figure 1.3).

1.4   POTENTIAL FAULTS IN A HYDRAULIC FLOW-INDUCED ROTARY SYSTEM

Different hydraulic flow-induced systems such as the Pelton turbine, Francis turbine, and Centrifugal pumps comprise many components such as buckets, shafts, rotor, nozzle, impeller, and bearing. Defects in these components could be broken impeller or bucket, clogging, added mass (scales), cavitation, misalignment in the shaft, and bearing defects. Some of the common defects are as follows.

Bucket defects: In Pelton turbines, the buckets are prone to damage, such as wavy erosion of the splitter. This erosion transforms the initially sharp splitter edge into a flattened, wavy surface curving inwards toward the runner's axis [20]. The wavy erosion pattern results from uneven wear along the splitter's length. Another type of erosion, ripple erosion, creates wave-like deformations across the bucket's curved surface, following the flow direction [21]. These ripples form from sediment particles sliding and scratching the bucket surface due to high acceleration in areas of low curvature. "Bulging erosion" describes a slight thinning and outward bulging of the flat area between the splitter and curved bucket surface [22]. Sediment scratching and sliding cause the bulging lines. In multi-jet Pelton turbines, combined cavitation and erosion near the bucket inlet and root, along with surface irregularities, generate secondary flows and splashing, leading to pitting. A polished surface with a metallic sheen can form around the runner due to the impact of small water droplets [23, 24]. Small sediment particles trapped in water droplets create an abrasive environment within the runner. Hydro-abrasive erosion of coated buckets depends on the coating's bond strength, properties, and application method. Hard coatings, being brittle, erode primarily on the splitter and cutout areas, with less erosion in the curved zone. **Seal defect**: The seal consists of two components: the inner diameter of the rotating seal element and the outer diameter of the stationary seal seat. Pump seals may experience failure or leakage due to extended periods of dry running, inappropriate lubrication (such as using heavy oil), excessive installation pressure, or damage incurred during the installation process. **Bearing fault**: The bearings are an integral part of the hydraulic machinery, which consists of different components such as inner race, outer race, and ball that are prone to defects. The bearing faces different defects such as brinelling, contamination, fretting, peeling and spalling etc. Brinelling refers to tiny localized indentation into the bearing race. Unlike brinelling, the small indentations (in the form of scratches, pitting, and scoring) are scattered on the bearing surface in case of contamination caused due to foreign fine particles, which are introduced through a defective bearing seal/lubricant. Fretting represents a flaw that arises from excessive friction between the inner race and the shaft. Peeling refers to a minor removal of the bearing surface typically less than 0.025 mm deep, primarily resulting from inadequate lubrication. Spalling occurs under metal fatigue and is an advanced stage of bearing defect. In this, a microscopic crack under the bearing surface makes its way to come to the surface, resulting in the flaking away of metal particles. **Impeller defect/Runner defects**: The impeller or runner of the centrifugal pump or Francis turbine is also subjected to the high impact of water and silt and corrosive material coming along with water. A clogged impeller occurs when silt or other particles block the impeller and get stuck together to the blade. Blade and Wheel cut occurs due to the erosive or abrasive action of the particles that are dissolved in the water or due to the corrosive or chemical action of the particles coming along with the water. Cavitation occurs when liquid pressure drops below its vapor pressure, creating vapor-filled cavities (bubbles) that collapse violently, generating damaging shock waves. A "missing blade" defect, caused by fatigue or metallurgical issues, involves a blade detaching from the impeller.

1.5 VIBRATION SIGNATURE OF FAULTS

Vibration signals contain a vast amount of data that can be categorized into various frequency ranges.

a. High-frequency zone:
   Surface roughness and corrosion on components (impeller, rotor, buckets, bearings) within the hydraulic rotary system cause high-frequency peaks in vibration spectra. These peaks reflect the increased stress and wear from the relative motion and contact between these components.
b. Natural frequency and defect frequency zones:
   There are impact events when one component makes relative movement with the defective areas that excite at its natural frequencies. In this scenario, the signal from defective machine element(s) has two components. The initial part of the signal is generated when one of the machine components interacts with the defective element, which is a low-frequency area that produces an amplitude peak within the fault frequency range. When any component strikes the trailing edge of the defective element, it results in an amplitude peak within the natural frequency range [25, 26]. The defect frequency generally depends on the defective elements.
c. Rotating frequency zone:
   Issues in machinery such as imbalance, bent shafts, misalignment, and looseness produce specific vibration patterns at the rotational speed and its multiples. These faults change the characteristics of contact and the distribution of load.

1.6 VIBRATION AND SIGNAL PROCESSING TECHNIQUES

Vibration waveform analysis is divided into three primary domains: time domain, frequency domain, and time-frequency domain techniques, as illustrated in Figure 1.4. The signals are analyzed within these three domains, and significant features are subsequently extracted for further examination.

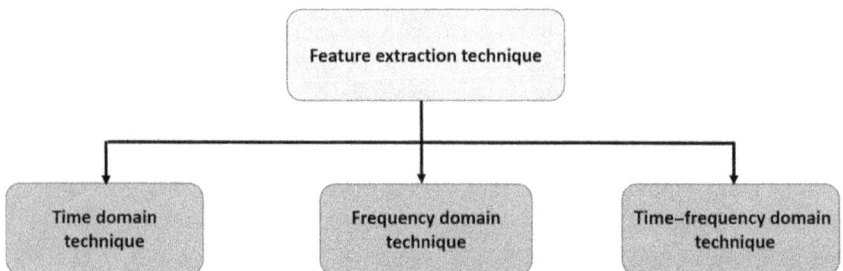

FIGURE 1.4 Vibration monitoring techniques.

1.6.1 Time-Domain Techniques

The time-domain analysis techniques in vibration signal processing are one of the simplest approaches that help in extracting the hidden information or features sensitive to defects. In this analysis, a series of digital waveforms are obtained that represent displacement, velocity, and acceleration.

1.6.2 Frequency-Domain Techniques

Frequency-domain analysis transforms digital waveforms to reveal their frequency components. This allows for easier identification and isolation of specific frequencies, a key advantage over time-domain analysis. The Fast Fourier Transform (FFT) is the most common method used.

1.6.3 Time–Frequency Domain Techniques

Frequency-domain analysis does not provide the same level of time resolution as time-domain analysis, resulting in a loss of time related information. Time-frequency analysis methods address this issue by examining signals in both time and frequency domains, making them well-suited for nonstationary signals. Some examples of these techniques are Short Time Fourier Transform (STFT), Wigner-Ville Distribution (WVD), Wavelet Transform (WT), Empirical Mode Decomposition (EMD), and Ensemble Empirical Mode Decomposition (EEMD) [27–31].

1.7 SIGNIFICANCE OF WORK

The harsh and complex operating conditions and different unpredictable operating factors affect the performance of the hydraulic flow-induced rotary systems and can result in various defects. These defects lead to severe damage in hydraulic flow machines and sometimes result in the shutdown of the whole plant. The sudden failure of different components not only results in an economic loss but is also a threat to life. Preventative and planned maintenance are part of the machine maintenance strategy, which is based on a set time interval and historical database. However, these strategies are ineffective because set maintenance operations either result in over or under-maintenance [32]. The CM approach used in this thesis not only detect different faults but also help in estimating the magnitude and severity of the defects, which can help maintenance personnel to determine when to shut down for replacement or maintenance. The entire procedure can be completed online without the need for any manual intervention or support.

1.8 SCOPE OF RESEARCH

The thesis mainly consists of the development of the fault identification scheme making use of vibration signal processing and artificial intelligence to detect different defects in hydraulic flow-induced rotary systems and their components.

In this book, different algorithms have been developed that classify the defects seeded on various components such as bearings, impellers, rotors, and buckets. Also, an attempt has been made to develop a vibration-based system through software and hardware to diagnose the defect in the pump.

Other defect diagnosis techniques, including acoustic emission, lubrication, and temperature analysis, are also available, however, they are beyond the scope of this study. The present study does not include stress analysis or fracture mechanisms.

1.9   ORGANIZATION OF THE BOOK

The book has been organized into six chapters.

Chapter 1 provides a brief overview of the techniques utilized in fault diagnosis of hydraulic flow-induced rotary systems. This chapter also includes a general summary of the book.

In **Chapter 2**, a system utilizing signal processing and artificial intelligence methods for diagnosing faults in Pelton turbines has been suggested. The efficiency of the proposed system has been demonstrated through experimental research and subsequently confirmed through comparative analysis.

In **Chapter 3**, a method for detecting defects on the inner race, outer race, and rollers of a Francis turbine bearing has been introduced. This method has been implemented on experimental data, and its effectiveness in diagnosing faults has been evaluated.

Chapter 4 focuses on the automated detection of defects in different parts of the centrifugal pump, including the bearing and impeller. The specifics of the experiments and the outcomes of the automated defect identification have been shared.

In **Chapter 5**, a method for identifying defects in the inner race, outer race, and roller of the bearing has been introduced. The method has been used on experimental signals, and its effectiveness in diagnosing faults is evaluated.

Finally, the scope of future research has been presented in **Chapter 6**.

2 Fault Diagnosis of the Pelton Turbine

2.1 INTRODUCTION

A Pelton turbine is an impulse turbine that generates mechanical energy by utilizing the momentum change of a fluid jet [33–35]. The rotor of a Pelton turbine, featuring buckets attached to a shaft and held in place by bearings, can be prone to harm due to factors such as corrosion, erosion, metallurgical flaws, cavitation, and insufficient lubrication [36, 37]. The primary components of a Pelton turbine that are susceptible to failure include bearings, nozzles, servomotors, and buckets. Condition monitoring, using acoustic or vibration analysis, can provide early warnings of defects. While acoustic monitoring offers advantages, vibration monitoring is often preferred due to its reduced sensitivity to environmental noise. This chapter focuses on two specific defect cases: bucket and bearing defects [38, 39].

2.2 DIAGNOSIS OF BUCKET DEFECT (CASE 1)

Pelton turbine buckets are susceptible to various forms of damage. Wavy erosion creates a wavy pattern on the splitter edge [20]; ripple erosion forms wave-like deformations on the curved surface due to sediment particle impact [21]. Bulging erosion involves thinning and outward bulging between the splitter and curved surface [22]. Cavitation erosion, particularly in multi-jet turbines, causes pitting, especially near the bucket inlet and root [23]. A polished surface can develop due to water droplet impact, trapping sediment and creating an abrasive environment [24]. Finally, hydro-abrasive erosion in coated buckets is influenced by coating quality and primarily affects the splitter and cutout areas.

Time-domain and frequency-domain analyses are widely used techniques for fault diagnosis [40]. Other techniques include wavelet transforms (WT) [40, 41], wavelet packet transforms (WPT) [42, 43], and signal sparse decomposition methods [44, 45]. However, WT and WPT are nonadaptive, requiring predefined wavelet functions [46, 47], while sparse decomposition, though adaptive, demands significant computational resources for large industrial datasets [48].

Ensemble empirical mode decomposition (EEMD) adaptively decomposes signals into intrinsic mode functions (IMFs) [49, 50], offering advantages in fault diagnosis despite limitations. A key weakness is the inability to separate components with frequencies within an octave, leading to mode mixing [51]. Additionally, empirical mode decomposition (EMD)'s susceptibility to noise (intermittence) [51], where a single IMF may contain multiple scales or similar scales appear

DOI: 10.1201/9781003614821-2

across multiple IMFs, also contributes to mode mixing. These issues have spurred the development of improved EMD methods.

EEMD, an improved version of EMD, addresses mode mixing by adding white noise before decomposition [52]. However, EEMD's limitations include difficulty in determining optimal white noise amplitude and ensemble number, hindering its adaptability. Furthermore, while EEMD significantly improves mode separation, it does not completely resolve the issue.

To reduce mode mixing in EMD, Li et al. [51] proposed a method called time-varying filtering-based EMD (TVF-EMD), which utilizes a B-spline approximation filter during the shifting process. TVF-EMD offers several advantages over existing methods such as EEMD and multivariate EMD [53]: (1) it simultaneously resolves separation and intermittence issues; (2) its time-varying filter handles mode mixing and time-varying features better than EEMD and variational mode decomposition; and (3) an improved stopping criterion enhances adaptability, especially at low sampling rates. Li et al. [51] demonstrated TVF-EMD's effectiveness through simulations and real-signal analysis, showing the significant influence of bandwidth threshold ξ (separation performance) and B-spline order n (filtering performance) [51]. However, optimal parameter selection (ξ and n) is crucial and challenging, requiring optimization techniques. The following contribution has been made in Case 1.

- An amended grey wolf optimization (AGWO) technique, which integrates position updating and Gaussian mutation strategies, is employed to determine ideal TVF-EMD parameters. This refinement helps the fundamental grey wolf optimization (GWO) algorithm avoid becoming stuck in local minima, thereby enhancing convergence speed and minimizing computational time.
- The suggested optimization algorithm employs kernel estimate for mutual information (KEMI) as its objective function. This function is integrated into the proposed health condition monitoring system for Pelton turbines.
- By utilizing the best parameters, TVF-EMD separates the original signal into multiple IMFs.
- The least KEMI value (fitness function) signifies the best solution, for which a scalogram is created for every health condition.
- Scalograms produced from the best solution are utilized to form training and testing datasets. A convolutional neural network (CNN) model, which is trained on this dataset, is assessed for classification accuracy using the testing set.

2.2.1 THEORETICAL BACKGROUND

2.2.1.1 Time-Varying Filter-Based Empirical Mode Decomposition (TVF-EMD)

EMD decomposes a signal $x(t)$ into IMFs and a residual $r(t)$, as shown in Eq. (2.1).

$$x(t) = \sum_{i=1}^{N} imf_i(t) + r(t) \tag{2.1}$$

where $imf_i(t)$ is ith IMF. In EMD, decomposition is a shifting process that takes place as per the given steps:

(1) Estimation of the "local mean" $m(t)$ and
(2) The mean $m(t)$ is recursively subtracted from the input signal until the stopping criteria are met.

In TVF-EMD, local narrow-band signals (with characteristics similar to mono-components, but yield improved Hilbert spectra) replace mono-components to enhance EMD performance. These signals are defined by an instantaneous bandwidth below a given threshold. The method involves determining local cut-off frequencies and applying a time-varying filter [51]. The shifting process in TVF-EMD uses this filter, following the steps outlined in [51].

A. Estimation of the local cut-off frequency
A B-spline approximation filter evaluates the dynamically changing cut-off frequency. This process includes generating polynomial splines that represent the input signal, as explained in Eq. (2.2).

$$g_m^n(t) = \sum_{k=-\infty}^{\infty} c(k)\beta^n(t/m - k) \tag{2.2}$$

The B-spline function, denoted as $\beta^n(t)$, along with coefficients c(k), order n, and knots m, defines the B-spline approximation. For given n and m, the approximation minimizes the squared error ε_m^2 by determining optimal coefficients $c(k)$.

$$\varepsilon_m^2 = \sum_{t=-\infty}^{+\infty} \left(x(t) - [c]_{\uparrow m} * b_m^n(t) \right)^2 \tag{2.3}$$

where $b_m^n(t) := \beta^n(t/m), [\bullet]_{\uparrow m}$ a m. The asterisk * indicates the convolution operation. Then $c(k)$ can be determined by

$$c(k) = \left[p_m^n * x \right]_{\downarrow m}(k) \tag{2.4}$$

In the above equation, $[\bullet]_{\downarrow m}$ represents down-sampling operation by m. The pre-filter is indicated by $p_m^n = \left[\left([b_m^n * b_m^n]_{\downarrow m} \right)^{-1} * b_m^n(t) \right]$. By substituting the value of $c(k)$, Eq. (2.2) takes the following form:

$$g_m^n = \left[p_m^n * x \right]_{\downarrow m} * b_m^n(t) \tag{2.5}$$

Eq. (2.5) defines a specific low-pass filter for B-spline approximation, where the knot spacing, m, influences the filter's local cut-off frequency. Because the knot information is initially unknown, the local cutoff frequency is first estimated from the input signal to construct the time-varying filter. This process proceeds as follows:

Step 1: Calculating the instantaneous amplitude, $A(t)$, and instantaneous frequency, $\varphi'(t)$, of a signal, $x(t)$, using the Hilbert transform.

$$A(t) = \sqrt{x^2 + \hat{x}(t)^2} \tag{2.6}$$

$$\varphi'(t) = d\left(\arctan\left(\hat{x}(t) / x(t)\right)\right) / dt \tag{2.7}$$

where $\hat{x}(t)$ is the Hilbert transform of $x(t)$ signal.

Step 2: Locate the maxima $\{t_{max}\}$ and minima $\{t_{min}\}$ of $A(t)$. The signal $z(t) = x(t) + j\hat{x}(x) = A(t)\exp\left(j\varphi(t)\right)$ is the analytical signal corresponding to $x(t)$. $\varphi(t)$ is the instantaneous phase represented as $\varphi(t) = \arctan\left[\hat{x}(t) / x(t)\right]$. In the case of a multicomponent signal, $z(t)$ is expressed as the combination of two signals.

$$z(t) = A(t)\exp\left(j\varphi(t)\right) = a_1 \exp\left(j\varphi_1(t)\right) + a_2 \exp\left(j\varphi_2(t)\right) \tag{2.8}$$

Thus, the following equations can be obtained:

$$A^2(t) = a_1^2(t) + a_2^2(t) + 2a_1(t)a_2(t)\cos\left[\varphi_1(t) - \varphi_2(t)\right] \tag{2.9}$$

$$\varphi'(t) = \frac{1}{A^2(t)}\left(\begin{array}{c} \varphi_1'(t)\left(a_1^2(t) + a_1(t)a_2(t)\cos\left[\varphi_1(t) - \varphi_2(t)\right]\right) \\ +\varphi_2'(t)\left(a_2^2(t) + a_1(t)a_2(t)\cos\left[\varphi_1(t) - \varphi_2(t)\right]\right) \end{array}\right) \\ + \frac{1}{A^2(t)}\left(\begin{array}{c} a_1'(t)a_2(t)\sin\left[\varphi_1(t) - \varphi_2(t)\right] \\ -a_2'a_1(t)\sin\left[\varphi_1(t) - \varphi_2(t)\right] \end{array}\right) \tag{2.10}$$

where $a_i(t)$ and $\varphi_i(t)$ represents amplitude and phase for the ith component. Using Eq. (2.9), the local minimum of $A(t)$ is determined at t_{min}, which satisfies the given equation:

$$\cos\left[\varphi_1(t_{min}) - \varphi_2(t_{min})\right] = -1 \tag{2.11}$$

On substituting Eq. (2.11) into Eq. (2.9) and Eq. (2.10), following Eqs. (2.12) and (2.13) are obtained:

$$A(t_{min}) = \left|a_1(t_{min}) - a_2(t_{min})\right| \tag{2.12}$$

$$\varphi'(t_{min}) A^2(t_{min}) = \varphi_1'(t_{min})\left(a_1^2(t_{min}) - a_1(t_{min})a_2(t_{min})\right)$$
$$+ \varphi_2'(t_{min})\left(a_2^2(t_{min}) - a_1(t_{min})a_2(t_{min})\right) \qquad (2.13)$$

Since $A(t_{min})$ is minima of $A(t)$, $A'(t_{min}) = 0$ is obtained. Thus,

$$a_1'(t_{min}) - a_2'(t_{min}) = 0 \qquad (2.14)$$

On solving Eqs. (2.11)–(2.14), $a_1(t_{min}), a_2(t_{min}), \varphi_1(t_{min})$, and $\varphi_2(t_{min})$ are computed. In similar way $a_1(t_{max}), a_2(t_{max}), \varphi_1(t_{max})$, and $\varphi_2(t_{max})$ are obtained from Eqs. (2.15)–(2.18).

$$\cos\left[\varphi_1(t_{max}) - \varphi_2(t_{max})\right] = 1 \qquad (2.15)$$

$$A(t_{max}) = a_1(t_{max}) + a_2(t_{max}) \qquad (2.16)$$

$$\varphi'(t_{max}) A^2(t_{max}) = \varphi_1'(t_{max})\left(a_1^2(t_{max}) + a_1(t_{max})a_2(t_{max})\right)$$
$$+ \varphi_2'(t_{max})\left(a_2^2(t_{max}) + a_1(t_{max})a_2(t_{max})\right) \qquad (2.17)$$

$$a_1'(t_{max}) + a_2'(t_{max}) = 0 \qquad (2.18)$$

Step 3: Computing $a_1(t)$ and $a_2(t)$.
The B-spline functions are given in Eqs. (2.19) and (2.20).

$$\beta_1(t) = |a_1(t) - a_2(t)| \qquad (2.19)$$

$$\beta_2(t) = |a_1(t) + a_2(t)| \qquad (2.20)$$

Using Eq. (4.9), the above equations can also be written as given in Eqs. (2.21) and (2.22):

$$\beta_1(t_{min}) = A(t_{min}) = |a_1(t_{min}) - a_2(t_{min})| \qquad (2.21)$$

$$\beta_2(t_{max}) = A(t_{max}) = |a_1(t_{max}) + a_2(t_{max})| \qquad (2.22)$$

Using interpolation technique between $A\{(t_{min})\}$ and $A\{(t_{max})\}$, the $\beta_1(t)$ and $\beta_2(t)$ can be easily computed. The $a_1(t)$ and $a_2(t)$ are slow varying components which can be computed using Eqs. (2.19) and (2.20). The modified form is given in Eqs. (2.23) and (2.24) as

$$a_1(t) = \left[\beta_1(t) + \beta_2(t)\right]/2 \qquad (2.23)$$

$$a_2(t) = \left[\beta_2(t) - \beta_1(t)\right] / 2 \qquad (2.24)$$

Step 4: Calculating φ_1' and φ_2'.

The $\eta(t)$ are expressed as given in Eqs. (2.25) and (2.26) which is the function of $\varphi_1'(t)$ and $\varphi_2'(t)$

$$\eta_1(t) = \varphi_1'\left[a_1^2(t) - a_1(t)a_2(t)\right] + \varphi_2'\left[a_2^2(t) - a_1(t)a_2(t)\right] \qquad (2.25)$$

$$\eta_2(t) = \varphi_1'\left[a_1^2(t) + a_1(t)a_2(t)\right] + \varphi_2'\left[a_2^2(t) + a_1(t)a_2(t)\right] \qquad (2.26)$$

Here, $a_1(t), a_2(t), \varphi_1'(t)$, and $\varphi_2'(t)$ are slow varying components. $\eta_1(t)$ and $\eta_2(t)$ are solved using Eqs. (2.25) and (2.26) by making interpolation between $\varphi'\left(\{t_{\min}\}\right)A^2\left(\{t_{\min}\}\right)$ and $\varphi'\left(\{t_{\max}\}\right)A^2\left(\{t_{\max}\}\right)$.

The components $\varphi_1'(t)$ and $\varphi_2'(t)$ take the form as given in Eqs. (2.27) and (2.28).

$$\varphi_1'(t) = \frac{\eta_1(t)}{2a_1^2(t) - 2a_1(t)a_2(t)} + \frac{\eta_2(t)}{2a_1^2(t) + 2a_1(t)a_2(t)} \qquad (2.27)$$

$$\varphi_2'(t) = \frac{\eta_1(t)}{2a_1^2(t) - 2a_1(t)a_2(t)} + \frac{\eta_2(t)}{2a_1^2(t) + 2a_1(t)a_2(t)} \qquad (2.28)$$

Using Eq. (2.10), Eq. (2.25) and Eq. (2.26) can be modified and written as given in Eqs. (2.29) and (2.30):

$$\eta_1(t_{\min}) = \varphi'(t_{\min})A^2(t_{\min}) = \varphi_1'(t_{\min})\left[a_1^2(t_{\min}) - a_1(t_{\min})a_2(t_{\min})\right]$$
$$+ \varphi_2'(t_{\min})\left[a_2^2(t_{\min}) - a_1(t_{\min})a_2(t_{\min})\right] \qquad (2.29)$$

$$\eta_2(t_{\max}) = \varphi'(t_{\max})A^2(t_{\max}) = \varphi_1'(t_{\max})\left[a_1^2(t_{\max}) - a_1(t_{\max})a_2(t_{\max})\right]$$
$$+ \varphi_2'(t_{\max})\left[a_2^2(t_{\max}) - a_1(t_{\max})a_2(t_{\max})\right] \qquad (2.30)$$

Step 5: Computing local cut-off frequency $\varphi_{bis}'(t)$. The local cut-off frequency is computed using Eq. (2.31) as given below.

$$\varphi_{bis}'(t) = \frac{\varphi_1'(t) + \varphi_2'(t)}{2} = \frac{\eta_2(t) - \eta_1(t)}{4a_1(t)a_2(t)} \qquad (2.31)$$

Step 6: Realigning $\varphi_{bis}'(t)$ to address the issue of intermittence problems.

Algorithm 1. Local cut-off frequency realignment
Step 1. Locate the maximum timing of $x(t)$, expressed as u_i, $i = 1,2,3...$
Step 2. Find out all intermittences, expressed as e_j, $j = 1,2,3...$, which satisfy $\dfrac{\max\left(\varphi'_{bis}(u_i:u_{i+1})\right)-\min\left(\varphi'_{bis}(u_i:u_{i+1})\right)}{\min\left(\varphi'_{bis}(u_i:u_{i+1})\right)} > \rho$ ($\rho =$ 0.25 is used in this work). Subsequently, the timing of u_i is taken as an intermittence, namely, $e_j = u_i$.
Step 3. For each e_j, if it is on rising edge (i.e., $\varphi'_{bis}(u_{i+1}) > \varphi'_{bis}(u_i)$), $\varphi'_{bis}(e_j: e_{j+1})$ is considered to be floor. If it is on falling edge (i.e., $\varphi'_{bis}(u_{i+1}) < \varphi'_{bis}(u_i)$), $\varphi'_{bis}(e_j: e_{j+1})$) is considered to be a floor. The remainders are regarded as peaks.
Step 4. Obtain the final local cut-off frequency by interpolating between the peaks.

FIGURE 2.1   Algorithm-1 for frequency realignment.

B. Filtering input signal to obtain local mean

Intermittent noise affects the local cutoff frequency, $\varphi'_{bis}(t)$, calculated in Step 5. Algorithm 1 [51] (Figure 2.1) addresses this using a time-varying filter to refine $\varphi'_{bis}(t)$.

With local cut-off frequency, the signal $h(t)$ can be viewed as in Eq. (2.32).

$$h(t) = \cos\left[\int \varphi'_{bis}(t)\,dt\right] \qquad (2.32)$$

A B-spline approximation filter is created using the extrema (knots) of $h(t)$, denoted as $\{t_{\min}\}$ and $\{t_{\max}\}$ of $h(t)$, to align the filter's cutoff frequency with $\varphi'_{bis}(t)$. This filter is then applied to the input signal, $x(t)$, resulting in an approximation, $m(t)$.

C. Verification of residual signal in meeting the stopping criterion

A narrowband signal is selected based on its instantaneous bandwidth. Eq. (2.33) expresses a relative criterion for this selection.

$$\theta(t) = \frac{B_{Loughlin}(t)}{\varphi_{avg}(t)} \qquad (2.33)$$

A signal is classified as narrowband if $\theta(t)$ is below the threshold $\xi\left[\theta(t) \le \xi\right]$. Eqs. (2.34) and (2.35), respectively, define the average instantaneous frequency $\varphi_{avg}(t)$ and the Loughlin instantaneous bandwidth $B_{Loughlin}$.

$$\varphi_{avg}(t) = \frac{a_1^2(t)\,\varphi'_1(t)+a_2^2(t)\,\varphi'_2(t)}{a_1^2(t)+a_2^2(t)} \qquad (2.34)$$

Algorithm 2. Shifting process of TVF-EMD

Step:1 Calculate $A(t)$ and $\varphi(t)$ using Eqs (4.6) and (4.7) respectively.
Step:2 Locate the local minima ($\{t_{min}\}$) and maxima ($\{t_{max}\}$) of $A(t)$ as per Eqs. (4.8) -(4.18)
Step:3 Calculate $a_1(t)$ and $a_2(t)$ using Eqs. (4.19) -(4.24).
Step:4 Calculate $\varphi_1'(t)$ and $\varphi_2'(t)$ according to Eqs. (4.25) -(4.30)
Step:5 Calculate $\varphi_{bis}'(t)$ through Eq. (4.31)
Step:6 Realign $\varphi_{bis}'(t)$ making use of Algorithm 1 to resolve the intermittence problem.
Step:7 Time varying filter is made to obtain the approximation result $m(t)$.
Step:8 Calculate $\theta(t)$ using Eqs. (4.33) -(4.35). If $\theta(t) \leq \xi$, $x(t)$ is taken to be an IMF. Else, let $x(t) - m(t)$ and repeat above steps.

FIGURE 2.2   Shifting process of TVF-EMD.

$$B_{Loughlin}(t) = \sqrt{\frac{a_1'^2(t) + a_2'^2(t)}{a_1^2(t) + a_2^2(t)} + \frac{a_1^2(t)a_2^2(t)\left(\varphi_1'(t) + \varphi_2'(t)\right)^2}{\left(a_1^2(t) + a_2^2(t)\right)^2}} \qquad (2.35)$$

The TVF-EMD shifting process steps are detailed in Algorithm 2 [51] (Figure 2.2).

D. Limitations of TVF-EMD

The bandwidth threshold, ξ, and B-spline order, n, which impact TVF-EMD performance (ξ affects separation, n affects filtering [51]), must be carefully chosen. Poor selection leads to mode mixing in the IMFs.

Finding the optimal combination of these parameters to best match the original signal is a key challenge addressed in this research.

2.2.1.2   Convolution Neural Network (CNN)

CNNs are deep learning tools that analyze images by processing and extracting information [54, 55]. In contrast to other types of networks, CNNs feature a three-dimensional structure of neurons. A standard CNN (Figure 2.3) typically consists of convolutional layers, a max-pooling layer, fully connected layers, and a classification layer.

A. **Convolution layer**: The convolutional layer, an essential part of CNNs, employs small filters that traverse the entire image through a process of shifting [56]. Convolution consists of computing the dot product of the filter with the image, aggregating the results over the area covered by the filter, and then continuing this procedure for the subsequent positions of the filter (Figure 2.3, Eq. 2.36).

$$z_j^l = \varphi\left(x_i^{l-1} * w_{ij}^{(1)l} + b_j^{(1)l}\right) \qquad (2.36)$$

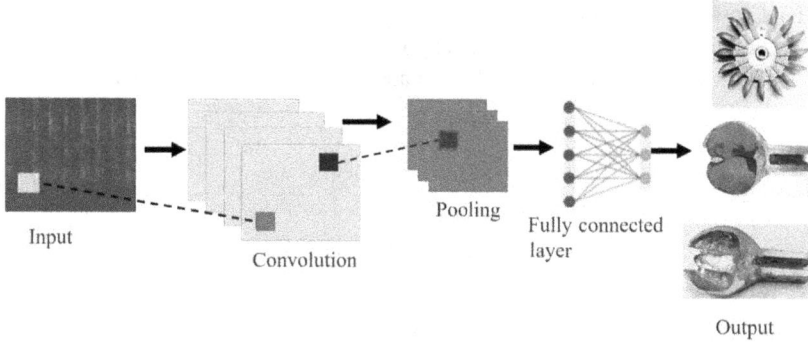

FIGURE 2.3 Convolution neural network architecture.

where φ represents the sigmoid activation function, $b_j^{(1)}$ is the bias at the lth layer, and $w_{ij}^{(1)}$ represents the weight between the ith input and j^{th} output at the lth layer.

B. **Pooling layer**: CNNs use pooling layers for downsampling. While various pooling functions exist, max pooling is commonly employed. The pooling expression is represented in Figure 2.3 and Eq. 2.37.

$$M_c^y(y) = \max\left(L_f^l(x)\right) for\ x = 1,1\ to\ pat_c, pat_w \qquad (2.37)$$

where $M_c^y(y)$ represents the pixel at location y in the lth convolutional layer for the cth channel, and pat_c and pat_w denote the patch height and width, respectively.

C. **Fully connected layer**: This layer is comparable to an artificial neural network (ANN), where neurons from earlier layers are linked to neurons in the following layers, as illustrated in Eq. 2.38.

$$y_j^l = \varphi\left(z_i^{l-1} * w_{ij}^{(2)l} + b_j^{(2)l}\right) \qquad (2.38)$$

where φ is the sigmoid function, $b_j^{(2)}$ is the bias, z_i^{l-1} is the input from the previous layer, and $w_{ij}^{(2)}$ represents the weight between the input and output nodes.

The fully connected layer, which connects the preceding layer's output to the next layer's neurons, involves a large number of training parameters.

D. **SoftMax layer**: This layer calculates the probability distribution over all potential target classes.

$$P\left(y_j^l\right) = \frac{\exp\left(y_j^l\right)}{\sum_{j=1}^{k} \exp\left(y_j^l\right)} \qquad (2.39)$$

(E) **Classification output layer**: This layer assesses the loss function through-out the training process, with the goal of reducing the cost function $\left(e^{existing}\right)$ of the CNN to enhance prediction accuracy, as illustrated in Eq. 2.40.

$$e^{existing} = CE + \psi \sum w^2 \tag{2.40}$$

where CE is cross-entropy loss as represented in Eq. (2.41).

$$CE = -\sum_{j=1}^{m} y_j^T \ln y_j^P \tag{2.41}$$

Here, y^P represents the predicted value, and the target value is y^T. The $L2$ regularization is represented by ψ.

2.2.1.3 Optimization of TVF-EMD using AGWO

At this stage, TVF-EMD parameter optimization (threshold and B-spline order) is performed to improve fault diagnosis from vibration signals. This optimization uses a search algorithm comprising a fitness function and a search method, detailed in the following section.

A. Kernel estimate for mutual information (KEMI)
KEMI quantifies the dependency between variables g and j, with high values indicating strong mutual information and zero indicating independence. Copula transformation (rank ordering) scales the variables to the $(0, 1)$ range before using Gaussian kernels to estimate marginal and joint probability distributions [57, 58]. This kernel estimator dynamically detects non-linear dependencies. For a dataset $z_i = \{g_i, j_i\}, i = 1, 2, \ldots, N$, the joint probability distribution, $p(z) = \frac{1}{N} \sum_i h^{-2} G\left(h^{-1}|z - z_i|\right)$, is computed, where G is the bivariate standard normal density, h is the kernel bandwidth, and N is the sample size. Marginal probabilities, $p(g)$ and $p(j)$, are derived from $p(z)$. KEMI is then calculated as shown in Eq. (2.42).

$$KEMI\left(g_i, j_i\right) = \frac{1}{N} \sum_i \log \frac{p\left(g_i, j_i\right)}{p\left(g_i\right)p\left(j_i\right)} \tag{2.42}$$

Table 2.1 presents the pseudo code for the KEMI calculation algorithm. To prevent mode mixing, the lowest KEMI value is chosen. In the decomposition process, the highest value is obtained from the cumulative information across all modes and the raw signal is utilized to extract pertinent information. Eq. (2.43) specifies the ratio of the overall mutual information among modes to the total mutual information between the original signal and its modes.

TABLE 2.1
Pseudocode to Calculate KEMI

Input: Variable g, Variable j
Output: KEMI (g, j)
1: Calculate the size of variables
2: Obtain Copula-transform
3: Calculate values for kernels at each data point
Kg = square form (exp (-ssd ([g; g])/h2)) + eye (Mg);
Kj = square form (exp (-ssd ([j; j])/h2)) + eye (Mj);
4: Calculate kernel sums for marginal probabilities
Cf = sum (Kg);
Cl = sum (Kj);
5: Kernel product for joint probabilities
Kgj = Kg. *Kj;
m = sum (Cg. *Cj) *sum (Kgj)./(Cg*Kj)./(Cj*Kg);
KEMI = mean (log(m));
ssd indicates the sum of squared differences

$$KEMI_{fitness\ function} = \frac{\sum_{K=1}^{K-1}KEMI\left(\mathrm{mod}(k),\mathrm{mod}(k+1)\right)}{\sum_{K=1}^{K}KEMI\left(\mathrm{mod}(k),g\right)} \qquad (2.43)$$

The KEMI fitness function reduces the mutual information among modes while enhancing the information derived from the original signal to avoid mode mixing.

B. Amended grey wolf optimization (AGWO)
  GWO, a meta-heuristic algorithm introduced by Mirjalili et al. [59], simulates wolf pack hunting. As shown in Figure 2.4, the wolf hierarchy is divided into four levels $(\alpha, \gamma, \delta, \omega)$. The top three agents (α, γ, δ) guide the optimization process, while the remaining wolves (ω) follow. Subsequent subsections detail GWO and AGWO, including the rationale behind AGWO's modifications.

- Encircling prey
  The initial stage of the hunting process, where wolves encircle their prey, is simulated by Eqs. (2.44) and (2.45).

$$D = \left| C.X_P(l) - X(l) \right| \qquad (2.44)$$

$$X(l+1) = X_p(l) - A.D \qquad (2.45)$$

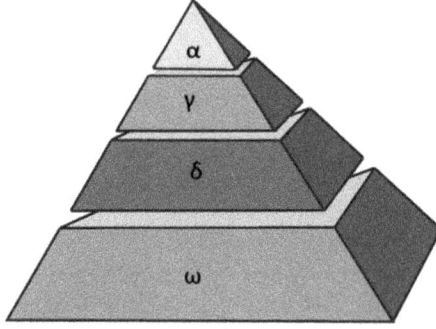

FIGURE 2.4   Hierarchy of grey wolf.

In these equations, l represents the current iteration, X_p denotes the prey's position, and X represents the grey wolf's position. The coefficient vectors A and C are calculated using Eqs. (2.46) and (2.47), respectively.

$$A = 2a.r_1 - a \tag{2.46}$$

$$C = 2.r_2 \tag{2.47}$$

where the random vectors r_1 and r_2 are within the interval [0, 1]. The parameter a is defined as $\left(2 - \left(\dfrac{2t}{T}\right)\right)$, where t is the current iteration and T is the maximum number of iterations. This ensures a linearly decreases from 2 to 0 over the iterations.

Grey wolf position updates, based on prey location, are calculated using Eqs. (2.44) and (2.45). This is achieved by adjusting the A and C vectors. The random vectors (wolves) r_1 and r_2 can assume any position, as detailed in Ref. [59].

- Hunting
  Initially, the top three search agents are randomly initialized, as the optimal solution (target) is unknown. Their positions are then saved and used to update the positions of the remaining agents. This process is described by Eqs. (2.48), (2.49), and (2.50).

$$D_\alpha = \left|C_1.X_\alpha - X\right|, D_\gamma = \left|C_2.X_\gamma - X\right|, D_\gamma = \left|C_3.X_\delta - X\right| \tag{2.48}$$

$$X_1 = X_\alpha - A_1.D_\alpha, X_2 = X_\gamma - A_2.D_\gamma, X_3 = X_\delta - A_3.D_\delta \tag{2.49}$$

$$X(t+1) = \frac{X_1 + X_2 + X_3}{3} \tag{2.50}$$

- Attacking prey
  The pursuit of the prey persists until it is caught, accomplished by lowering the value of a in the hunting process framework. Lowering a also lessens the variations in A (a random value ranging from $\left[-2a, 2a\right]$). When $|A|$ is less than 1, the wolves gather around the prey.
- Search for prey (exploration)
  Exploration is crucial for effective hunting; therefore, AGWO incorporates additional operators during exploration to enhance performance. Grey wolves initially disperse while searching for prey, converging only upon locating it to initiate an attack.

Mathematically, $|A| \geq 1$ causes the grey wolves to diverge, searching for better prey. Simultaneously, C takes random values to emphasize exploration, regardless of iteration number. C also weights each prey, guiding the wolves toward it.

2.2.1.4 Proposed Modifications in GWO

This section outlines the enhancements made to the fundamental GWO algorithm. These upgrades, which involve a position-updating method and a Gaussian mutation approach centered on population division and reconstruction, are designed to improve the search capability of the algorithm.

A. Position updating
  During the hunt, wolf positions are updated according to a normal distribution. Eq. (2.51) incorporates weighting factors V_1, V_2, and V_3 into Eq. (2.50) to achieve this position update.

$$X(t+1) = \frac{V_1 X_1 + V_2 X_2 + V_3 X_3}{3} \tag{2.51}$$

where $V_1 + V_2 + V_3 = 1$, and $V_1, V_2,$ and V_3 are drawn from a normal distribution. The ranges for V_1 and V_2 were determined empirically through experimentation to optimize performance.

$$V_1 = \left[0.1, 0.5\right] \tag{2.52}$$

$$V_2 = \left[0.7, 0.1\right] \tag{2.53}$$

$$V_3 = 1 - \left[V_1 + V_2\right] \tag{2.54}$$

In this context, 0.1 represents the mean and 0.5 signifies the variance for V_1. Likewise, 0.7 denotes the mean and 0.1 indicates the variance for V_2 in order to identify the optimal solution.

B. Gaussian mutation strategy

Back and Schwefel [60] presented a Gaussian mutation approach to enhance the effectiveness of metaheuristic algorithms [61, 62]. This method creates new solutions that are close to current ones by taking small steps to navigate the search space while preserving diversity. The Gaussian density function is characterized as follows:

$$f_{0,\sigma^2}(\theta) = \frac{1}{\sigma\sqrt{(2\pi)}} e^{\frac{-\theta^2}{2\sigma^2}}$$ (2.55)

In this context, σ^2 denotes the variance associated with the possible solution. Eq. (2.55) illustrates a probability density function that has a minimal value at a mean of 0 and a standard deviation of 1. A random vector, $Gaussian(\theta)$, is generated using this density function as shown in Eq. 2.56.

$$G_i' = G_i\left(1 + Gaussian(\theta)\right)$$ (2.56)

where G_i is ith mutated candidate solution and $Gaussian(\theta)$ is a random vector.

2.2.2 DEFECT IDENTIFICATION SCHEME

An AGWO algorithm is proposed to determine the optimal bandwidth threshold (ξ) and B-spline order (n) for TVF-EMD, using KEMI as the fitness function [Eq. 2.57]. This enhanced GWO algorithm incorporates position updating and Gaussian mutation strategies to minimize the fitness function as presented in Eq. 2.57.

$$\begin{cases} objective\ function = \min_{(\gamma = \{\xi,n\})}(KEMI) \\ s.t, \xi\ \varepsilon\left[0,0.8\right] \\ n = 5,6,\ldots,30 \end{cases}$$ (2.57)

Here, KEMI denotes the kernel estimate of mutual information utilized in the TVF-EMD decomposition modes. The parameters $\gamma = (\xi,n)$ refer to the TVF-EMD parameters that need to be optimized, where ξ (bandwidth threshold) varies between [0, 0.8] and n (B-spline order) ranges from [5, 30], as indicated in the literature [54]. The procedure for applying the proposed method is outlined in the following steps:

Step 1: Collect vibration data from the Pelton turbine. Set up the AGWO algorithm with a specified population size N and a maximum number of iterations L, applying a defined range for the TVF-EMD parameters. Document the objective function value for every iteration.

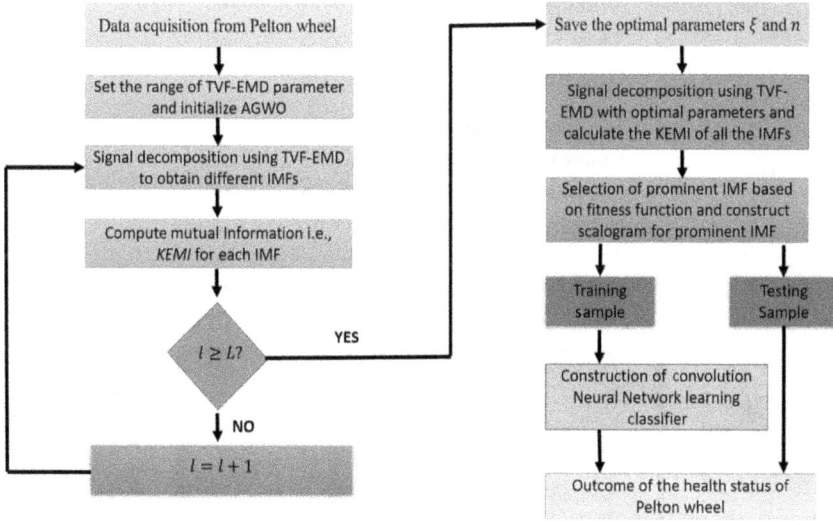

FIGURE 2.5   Fault detection methodology.

Step 2: Break down the raw vibration data into IMFs utilizing TVF-EMD and compute the KEMI for all IMFs. Document the lowest objective function value for every iteration.

Step 3: When the iteration count l meets or exceeds the maximum allowed iterations L, the algorithm stops running. If not, increase l and keep iterating.

Step 4: The combination of parameters that produces the lowest value for the KEMI fitness function is determined and recorded. The IMF associated with this lowest fitness value is referred to as the sensitive IMF.

Step 5: Scalograms are produced for the sensitive IMF and saved to form image data.

Step 6: The image data is fed into a CNN to assess classification accuracy. The process for fault diagnosis is demonstrated in Figure 2.5.

2.2.3   APPLICATION OF DEFECT IDENTIFICATION SCHEME ON PELTON TURBINE

A. Test rig

The suggested fault diagnosis method is implemented on a Pelton turbine (refer to Table 2.2 for specifications and Figure 2.6 for the test rig). This turbine consists of components such as a rotor with 16 buckets, casing, and nozzle, which are prone to defects. The buckets function as cantilever beams, directly experiencing the force of the water jet, which leads to fatigue. The rotor is mounted on a shaft and is supported by two SKF UC 206 bearings, driven by a 15 hp motor combined with a centrifugal pump system. The buckets receive the impact of the tangential water jet emitted from the nozzle.

TABLE 2.2
Specification of Pelton Turbine

Maximum Output	3 KW
Supply head	30 m
Maximum discharge	400 liter per minute
Sump tank capacity	200 liters
Number of buckets	16

Uni-axial accelerometer

Side view of pelton wheel

Data logging unit

Lab view interface

Data acquisition system

FIGURE 2.6   Pelton turbine test rig.

(B) Data acquisition

Vibration data was gathered for four different conditions of the Pelton turbine: healthy, splitter wear, added mass, and a missing bucket (see Figure
2.7). For each of these conditions, 350 signals were collected at seven distinct speeds (ranging from 900 to 1500 rpm) using a uniaxial accelerometer from PCB® Piezotronics (with a sensitivity of 100 mV/g) installed on a
bearing. A National Instruments data acquisition system with 24-bit resolution and 4 channels, operating within a LabVIEW environment, recorded
the data at a frequency of 70,000 Hz, resulting in 14,000 data points for
each sample. The raw signals underwent processing using an optimally
parameterized TVF-EMD to extract IMFs. The mode most sensitive to
detection (as established by KEMI) was chosen to create continuous wavelet transform scalograms, thus producing the image dataset. The analysis
was performed using MATLAB R2019a, with data acquisition conducted
through LabVIEW 2020. The specifications of the system used were:
Intel(R) Core(TM) i5-4210U CPU running at 1.70 GHz and capable of
2.40 GHz, equipped with 8 GB of RAM, and operating on 64-bit
Windows 10.

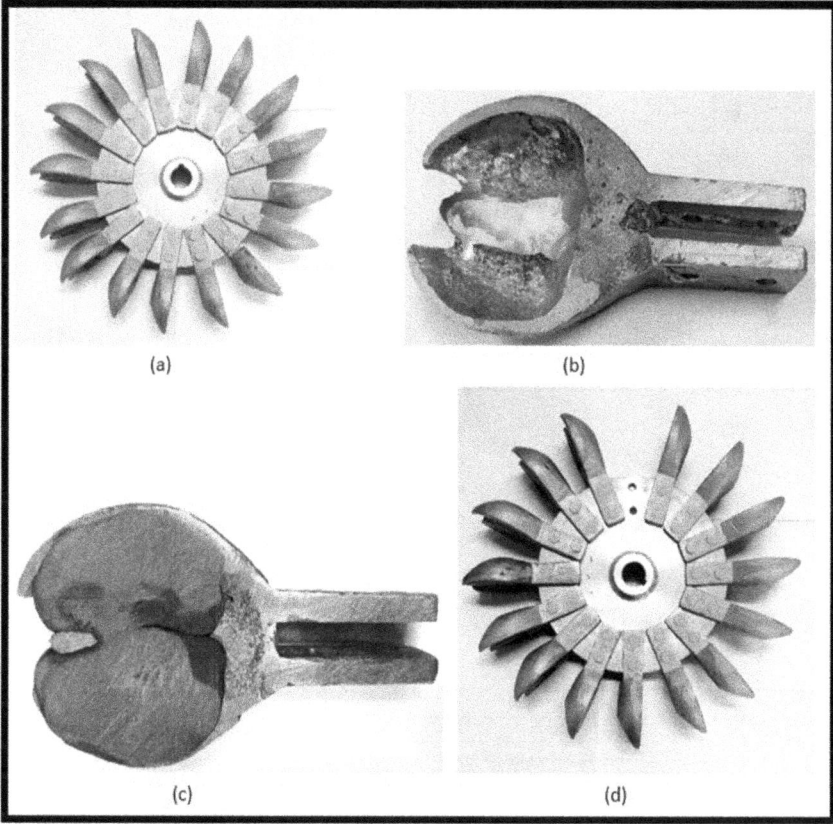

FIGURE 2.7 Various health states: (a) optimal health, (b) wear on the splitter, (c) additional mass, and (d) absence of one bucket.

Figure 2.8(a) shows a typical time-domain vibration signal at 1200 rpm, while Figure 2.8(b) displays its TVF-EMD decomposition into IMFs using the optimized parameters $\left(\xi = 0.0073, n = 23 \right)$ obtained via the proposed AGWO algorithm. KEMI values were calculated for each IMF: 0.0379, 0.0338, 0.0360, 0.0313, 0.0331, 0.0337, 0.0302, 0.0375, 0.0365, and 1.8826. IMF 7, exhibiting the minimum KEMI value (and thus minimum mutual information with the raw signal), was selected for further analysis. Its scalogram is shown in Figure 2.8(c).

Splitter wear was simulated by grinding approximately 70% of a bucket's splitter. Figure 2.9(a) shows a raw vibration signal under this condition. AGWO optimization, using KEMI, yielded optimal TVF-EMD parameters of $\xi = 0.0313$ and $n = 18$ at 1200 rpm. The resulting decomposition is shown in Figure 2.9(b). KEMI values for the 10 IMFs were: 0.0350, 0.0346, 0.0346, 0.0405, 0.0317, 0.0324, 0.0351, 0.342, 0.0349, and 1.7553. IMF 5, exhibiting the least mutual information, was selected for scalogram generation (Figure 2.9c).

Figure 2.10(a) shows a typical vibration signal at 1200 rpm with added mass (25 grams) on a bucket to simulate imbalance. The TVF-EMD

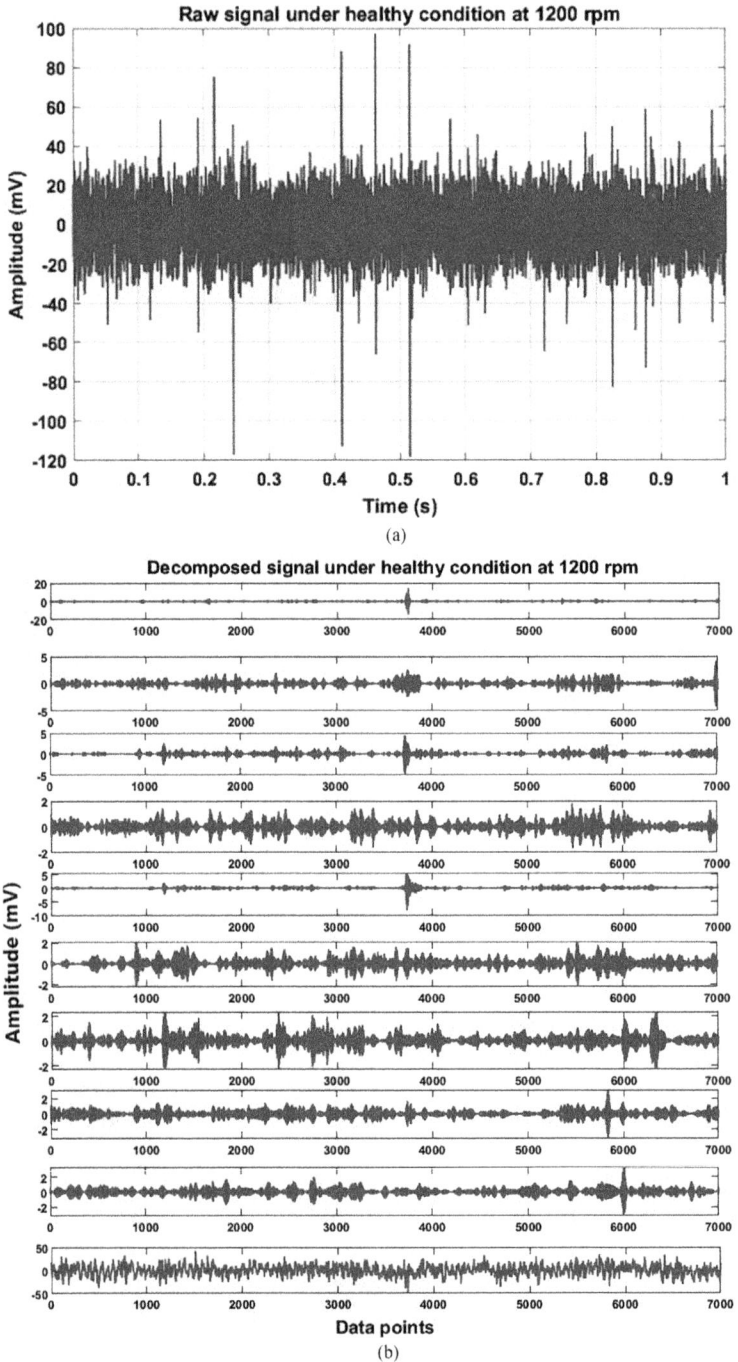

FIGURE 2.8 (a) Raw signal, (b) decomposed signals, and

(*Continued*)

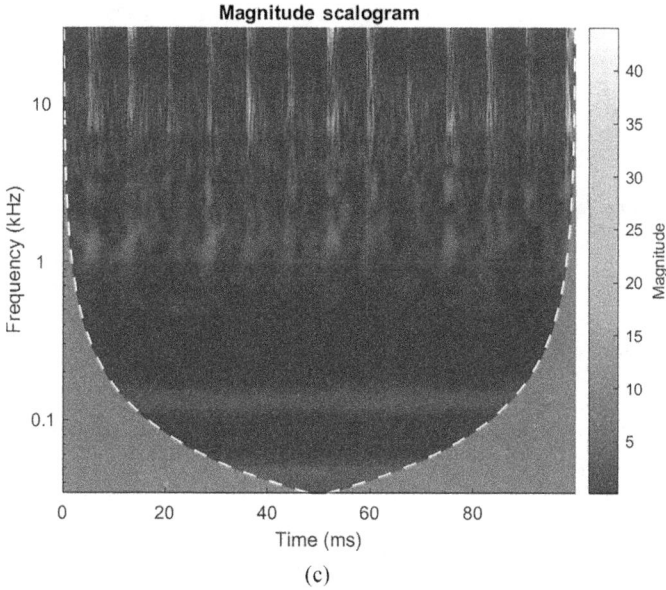

FIGURE 2.8   (Continued) (c) scalogram at 1200 rpm under healthy condition.

decomposition is shown in Figure 2.10(b). Using the previously determined optimal TVF-EMD parameters, KEMI values for the IMFs were calculated: 0.0365, 0.0387, 0.0323, 0.0321, 0.0367, 0.0343, 0.0362, 0.0357, 0.0336, and 2.0414. IMF 4, possessing the minimum KEMI value (and therefore minimum

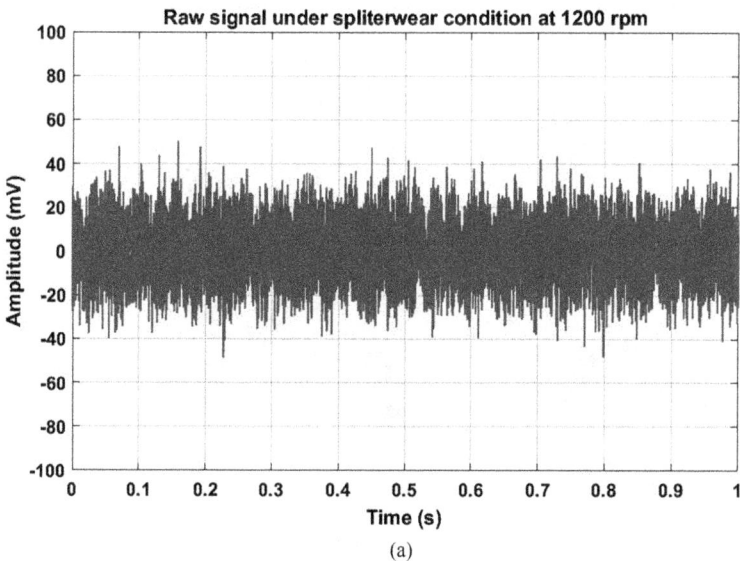

FIGURE 2.9   (a) Unprocessed signal,

(Continued)

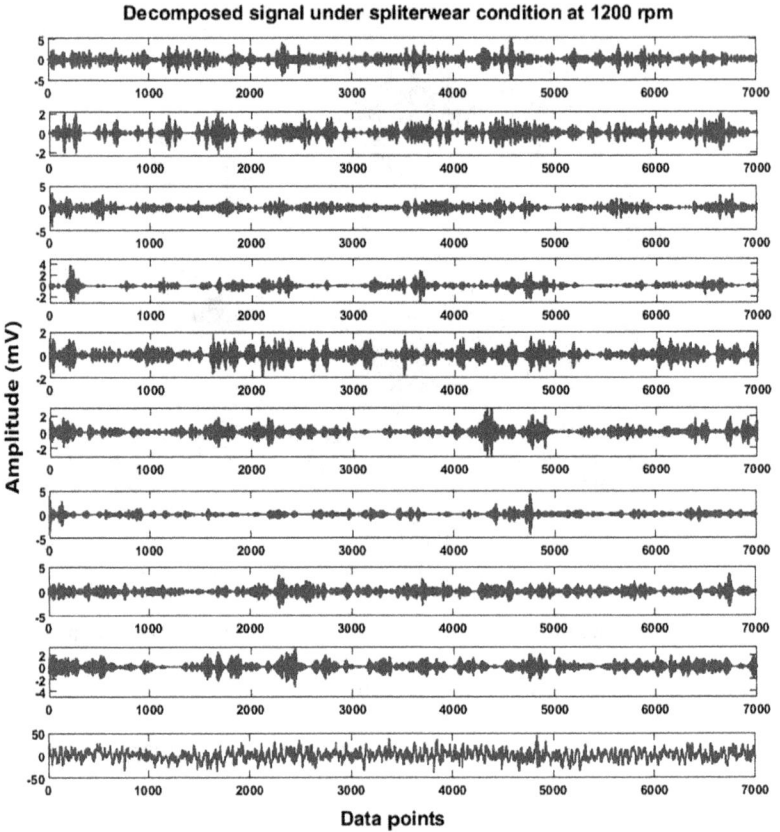

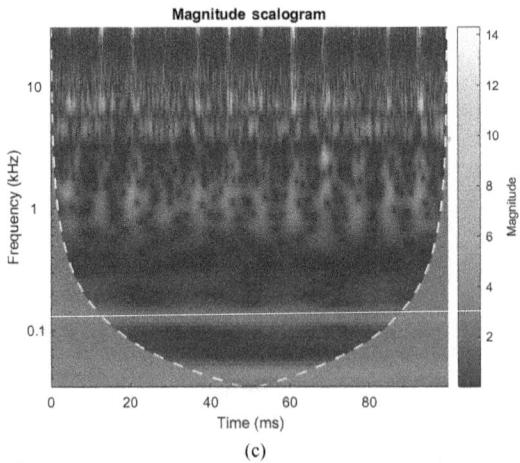

FIGURE 2.9 **(Continued)** (b) decomposed signals, and (c) scalogram at 1200 rpm under conditions of splitter wear.

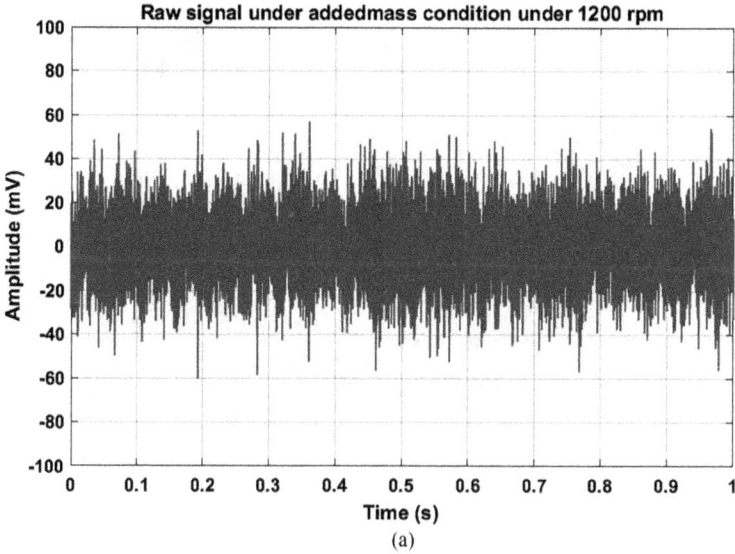

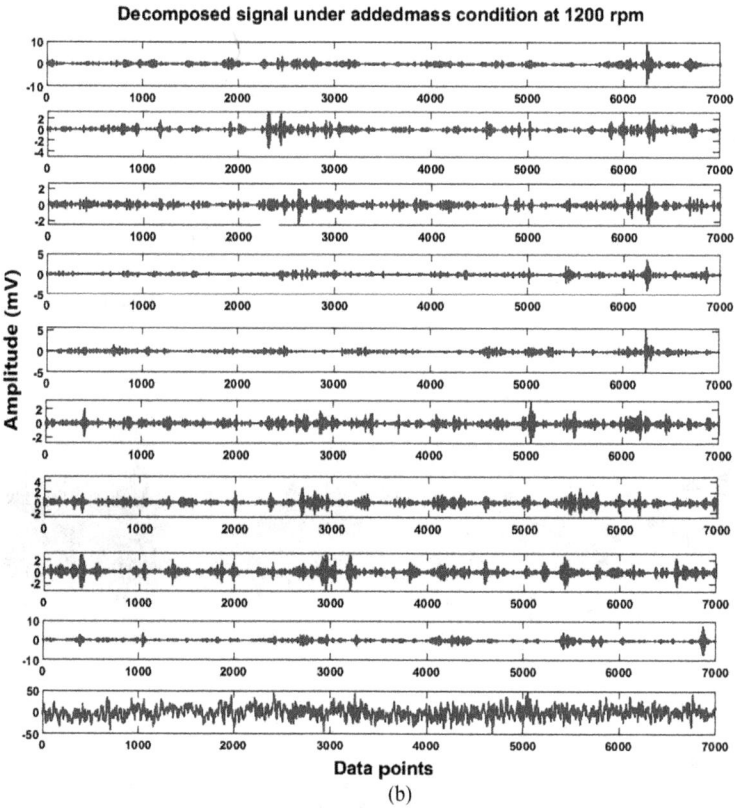

FIGURE 2.10 (a) Unprocessed signal, (b) decomposed signals,

(*Continued*)

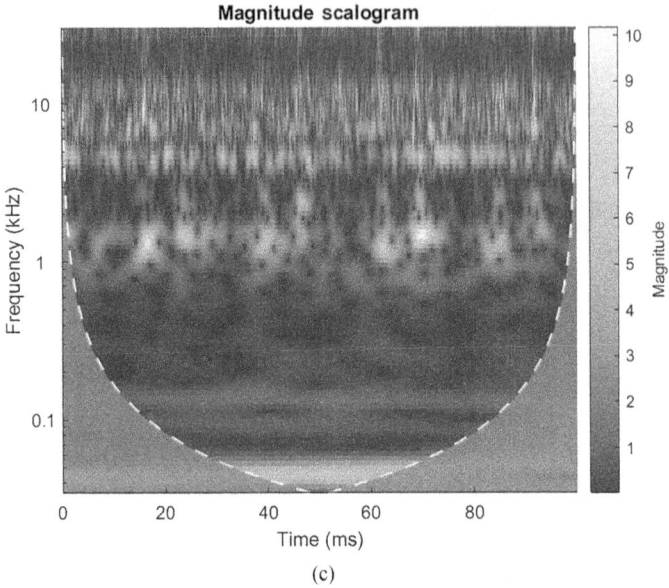

Magnitude scalogram

(c)

FIGURE 2.10 **(Continued)** (c) scalogram at 1200 rpm with added mass condition.

mutual information with the raw signal), was selected for scalogram generation (Figure 2.10c).

Figure 2.11(a) shows a typical 1200 rpm vibration signal with a missing bucket (simulating imbalance). Optimized TVF-EMD $\left(\xi = 0.0095, n = 19\right)$

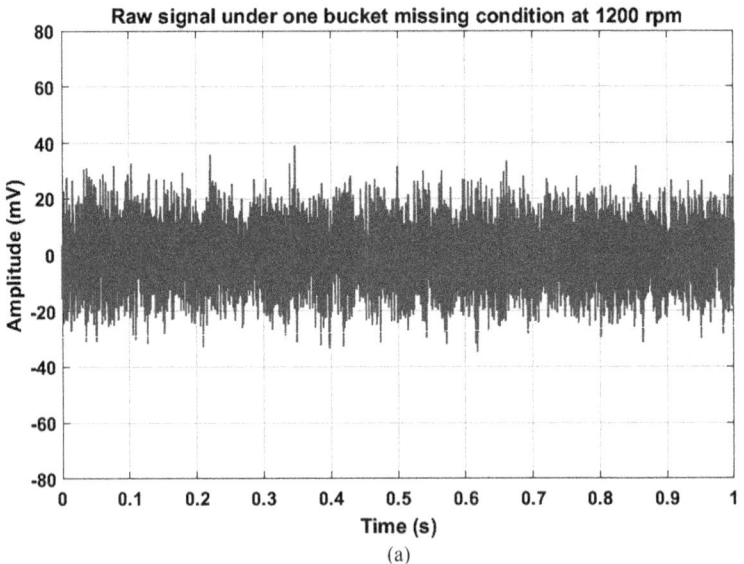

FIGURE 2.11 (a) Unprocessed signal,

(Continued)

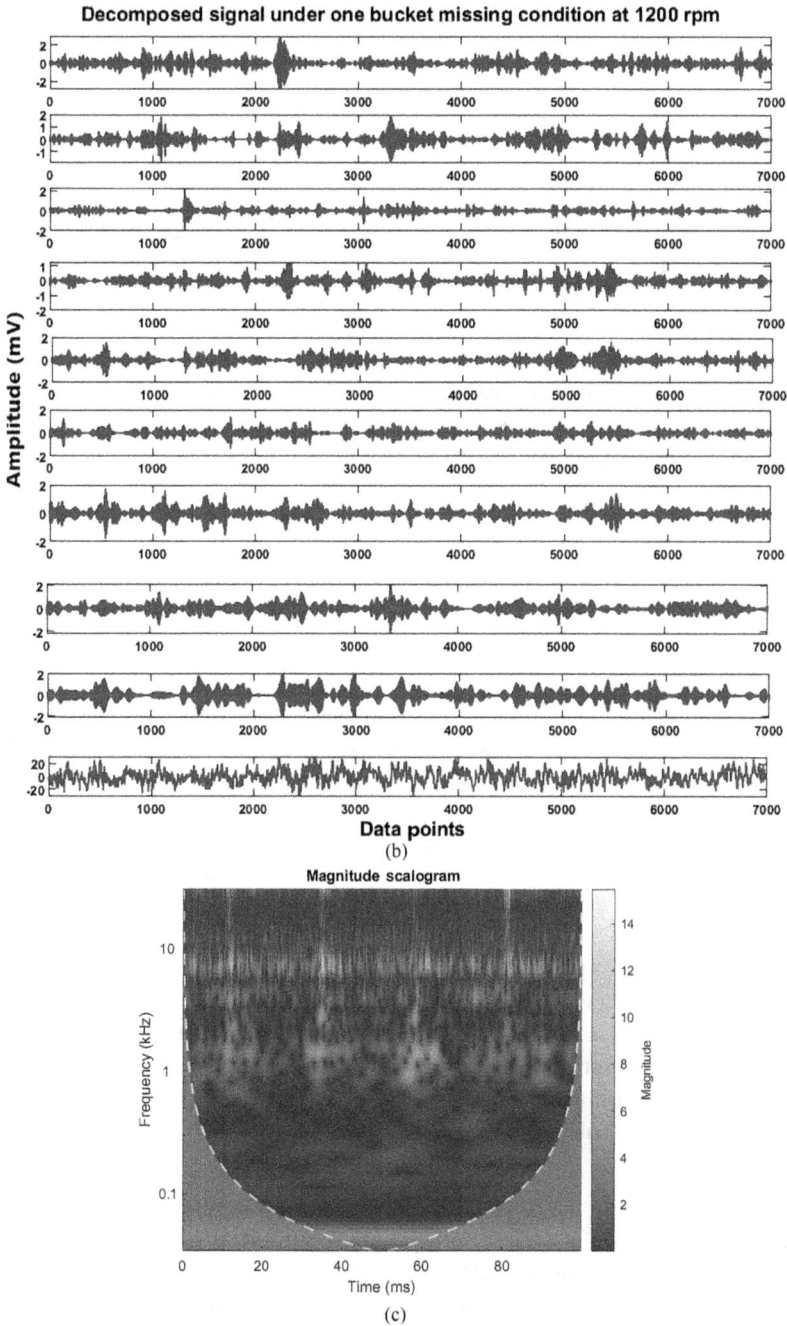

FIGURE 2.11 **(Continued)** (b) decomposed signals, and (c) scalogram recorded at a speed of 1200 rpm with one bucket absent.

decomposes this signal into ten IMFs [Figure 2.11(b)]. IMF 6, selected based on the minimum KEMI value, is used to generate the scalogram as shown in Figure 2.11(c).

2.2.4 RESULTS AND DISCUSSION

2.2.4.1 Comparison of the AGWO with Other Art of Optimization

The AGWO algorithm was evaluated using 23 benchmark functions (see Table 2.3) and was compared to GWO, salp swarm algorithm (SSA), sine-cosine (SCA), whale optimization algorithm (WOA), ant lion optimization (ALO), and grasshopper optimization algorithm (GOA) based on their mean results and standard deviations (refer to Table 2.4). AGWO achieved better performance than the other algorithms on 17 of the functions (F1–F7, F9–F11, F15, F17–F18, F20–F22, and F23), showing the smallest standard deviation. SCA excelled on functions F8 and F16, while GOA was the top performer on F12 and F13, SSA on F14, and ALO on F19. These findings highlight the exceptional optimization capabilities of AGWO.

Here, C denotes the function's characteristics, and D represents the dimensionality. US, UN, MS, and MN refer to unimodal separable, unimodal non-separable, multimodal separable, and multimodal non-separable functions, respectively.

The algorithms were evaluated on benchmark functions by calculating the mean and standard deviation over 20 independent runs (refer to Table 2.5), but individual run comparisons were not made. To determine statistical significance, a Wilcoxon rank-sum test was conducted at a 5% significance level, resulting in P-values (see Table 2.5). P-values under 0.05 lead to the rejection of the null hypothesis, suggesting there are statistically significant differences. The algorithm that performed best (based on the lowest standard deviation) for each function was compared to the others. Instances of self-comparison are indicated as N/A.

Table 2.5 shows AGWO achieved the best results for 15 functions (F1–F7, F9–F11, F17, F19, and F21–F23), while SCA performed best on F8, F15, F16, and F18; GOA on F12, F13, and F20; and GOA and SSA performed equally well on F14. Tables 2.4 and 2.5 demonstrate AGWO's statistically significant superiority over other algorithms [60], indicating its enhanced effectiveness for Pelton turbine fault identification.

2.2.4.2 The Need for Optimizing TVF-EMD Parameters

The TVF-EMD technique breaks down signals into IMFs, selecting a significant IMF through an appropriate index. Typically, the parameters of TVF-EMD (ξ and n) are determined based on empirical methods, which may impact the accuracy of the decomposition. This study suggests utilizing AGWO, an enhanced

TABLE 2.3
Definition of Benchmark Functions

S. No.	Function	Formulation	D	Range	C	Global Min.
F1	Sphere	$F(x) = \sum_{i=1}^{D} x_i^2$	30	[−100, 100]	US	0
F2	Schwefel 2.22	$F(x) = \sum_{i=1}^{D}|x_i| + \prod_{i=1}^{D}|x_i|$	30	[−10, 10]	UN	0
F3	Schwefel 1.2	$F(x) = \sum_{i=1}^{D}\left(\sum_{j=1}^{i} x_j^2\right)$	30	[−100, 100]	UN	0
F4	Schwefel 2.21	$F(x) = \max_i\{|x_i|, 1 \le i \le D\}$	30	[−100, 100]	US	0
F5	Rosenbrock	$F(x) = \sum_{i=1}^{D-1}\left[100\left(x_{i+1} - x_i^2\right)^2 + (x_i - 1)^2\right]$	30	[−30, 30]	UN	0
F6	Step	$F(x) = \sum_{i=1}^{D}\left(\lfloor x_i + 0.5\rfloor\right)^2$	30	[−100, 100]	US	0
F7	Quartic	$F(x) = \sum_{i=1}^{D} i x_i^4 + random[0,1]$	30	[−1.28, 1.28]	US	0
F8	Schwefel	$F(x) = -\sum_{i=1}^{D}\left(x_i \sin\sqrt{|x_i|}\right)$	10	[−500, 500]	MS	−418.9829*D
F9	Rastrigin	$F(x) = 10D + \sum_{i=1}^{D}\left(x_i^2 - 10\cos\left(2\pi x_i\right)\right)$	30	[−5.12, 5.12]	MS	0

(Continued)

TABLE 2.3 (CONTINUED)

Definition of Benchmark Functions

S. No.	Function	Formulation	D	Range	C	Global Min.
F10	Ackley	$F(x) = -20\exp\left(-0.2\sqrt{\dfrac{1}{D}\sum_{i=1}^{D}x_i^2}\right) - \exp\left(\dfrac{1}{D}\sum_{i=1}^{D}\cos(2\pi x_i)\right) + 20 + e$	30	[−32, 32]	MN	0
F11	Griewank	$F(x) = \dfrac{1}{4000}\sum_{i=1}^{D}x_i^2 - \prod_{i=1}^{D}\cos\left(\dfrac{x_i}{\sqrt{i}}\right) + 1$	30	[−600, 600]	MN	0
F12	Penalized	$F(x) = \dfrac{\pi}{D}\left\{10\sin^2(\pi y_1) + \sum_{i=1}^{D-1}(y_i-1)^2\left[1+10\sin^2(\pi y_{i+1})\right] + (y_D-1)^2\right\}$ $+ \sum_{i=1}^{D}u(x_i,10,100,4)$ where $y_i = 1 + \dfrac{x_i+1}{4}$, and $u(x_i,a,k,m) = \begin{cases} k(x_i-a)^m & ;x_i > a \\ 0, & -a < x_i < a \\ k(-x_i-a)^m & ;x_i < -a \end{cases}$	30	[−50, 50]	MN	0
F13	Penalized 2	$F(x) = 0.1\left\{\sin^2(3\pi x_1) + \sum_{i=1}^{D}(x_i-1)^2\left[1+\sin^2(3\pi x_{i+1})\right] + (x_D-1)^2\left[1+\sin^2(2\pi x_D)\right]\right\} + \sum_{i=1}^{D}u(x_i,5,100,4)$	30	[−50, 50]	MN	0
F14	Foxholes	$F(x) = \left[\dfrac{1}{500} + \sum_{j=1}^{25}\dfrac{1}{j+\sum_{i=1}^{D}(x_i-a_{ij})^6}\right]$	2	[−65.536, 65.536]	MS	0.998004

ID	Name	Function		Range		Value
F15	Kowalik	$F(x) = \sum_{i=1}^{11}\left[a_i - \dfrac{x_1\left(b_i^2 + b_i x_2\right)}{b_i^2 + b_i x_3 + x_4}\right]^2$	4	[−5, 5]	MN	0.0003075
F16	Six-hump camel-back	$F(x) = 4x_1^2 - 2.1x_1^4 + \dfrac{1}{3}x_1^6 + x_1 x_2 - 4x_2^2 + 4x_2^4$	2	[−5, 5]	MN	−1.0316285
F17	Branin	$F(x) = \left(x_2 - \dfrac{5.1}{4\pi^2}x_1^2 + \dfrac{5}{\pi}x_1 - 6\right)^2 + 10\left(1 - \dfrac{1}{8\pi}\right)\cos x_1 + 10$	2	[−5, 5]	MN	0.398
F18	Goldstein-Price	$F(x) = \left[1 + (x_1 + x_2 + 1)^2\left(19 - 14x_1 + 3x_1^2 - 14x_2 + 6x_1 x_2 + 3x_2^2\right)\right]$ $\times \left[30 + (2x_1 - 3x_2)^2 \times \left(18 - 32x_1 + 12x_1^2 + 48x_2 - 36x_1 x_2 + 27x_2^2\right)\right]$	2	[−2, 2]	MN	3
F19	Hartman 3	$F(x) = -\sum_{i=1}^{4} c_i \exp\left[-\sum_{j=1}^{3} a_{ij}\left(x_j - p_{ij}\right)^2\right]$	3	[−5, 5]	MN	−3.862782
F20	Hartman 6	$F(x) = -\sum_{i=1}^{4} c_i \exp\left[-\sum_{j=1}^{6} a_{ij}\left(x_j - p_{ij}\right)^2\right]$	6	[−5, 5]	MN	−3.32236
F21	Shekel5	$F(x) = -\sum_{i=1}^{5}\left[(x - a_i)(x - a_i)^T + c_i\right]^{-1}$	4	[−5, 5]	MN	−10.1532
F22	Shekel7	$F(x) = -\sum_{i=1}^{7}\left[(x - a_i)(x - a_i)^T + c_i\right]^{-1}$	4	[−5, 5]	MN	−10.4029
F23	Shekel10	$F(x) = -\sum_{i=1}^{10}\left[(x - a_i)(x - a_i)^T + c_i\right]^{-1}$	4	[−5, 5]	MN	−10.5364

TABLE 2.4

Comparison of the Proposed Algorithm with Other State of Art

Benchmark Function		Optimizations Algorithms						
S. No	Parameters	AGWO (Proposed)	GWO	SSA	SCA	WOA	ALO	GOA
F1	Average	0.0000	1.6822×10^{-27}	9.2785×10^{-09}	5.8635×10^{-28}	1.0749×10^{-174}	1.5679×10^{-09}	1.5789×10^{-09}
	Std	**0.0000**	3.2926×10^{-27}	2.5275×10^{-09}	2.5965×10^{-27}	0.0001	6.1292×10^{-10}	1.2995×10^{-09}
F2	Average	6.1135×10^{-229}	7.4670×10^{-17}	0.0177	2.3927×10^{-21}	3.1655×10^{-107}	1.6047×10^{-04}	0.9994
	Std	**0.0000**	5.5843×10^{-17}	0.0795	5.1292×10^{-21}	1.4178×10^{-106}	4.3480×10^{-04}	1.5428
F3	Average	0.0000	2.4797×10^{-05}	1.4655×10^{-09}	1.3645×10^{-11}	$1.2419 \times 10^{+04}$	6.3561×10^{-07}	1.2175×10^{-07}
	Std	**0.0000**	7.4078×10^{-05}	6.8216×10^{-10}	4.1399×10^{-11}	$7.9873 \times 10^{+03}$	9.7641×10^{-07}	2.8859×10^{-07}
F4	Average	1.3248×10^{-217}	7.1546×10^{-07}	1.3318×10^{-05}	5.9065×10^{-10}	32.2856	4.9145×10^{-05}	2.6025×10^{-05}
	Std	**0.0000**	8.1491×10^{-07}	2.4181×10^{-06}	1.7091×10^{-09}	28.5950	3.7018×10^{-05}	1.3877×10^{-05}
F5	Average	0.0000	27.8866	42.1985	6.9228	26.5109	62.2978	78.1317
	Std	**0.0000**	0.7314	71.3154	0.3861	0.3158	79.7189	258.6871
F6	Average	0.0000	0.7193	6.3532×10^{-10}	0.3368	0.0140	1.8139×10^{-09}	9.9609×10^{-10}
	Std	**0.0000**	0.3558	2.7101×10^{-10}	0.1565	0.0118	7.6227×10^{-10}	6.0218×10^{-10}
F7	Average	6.5921×10^{-05}	0.0018	0.0058	0.0008	8.7925×10^{-04}	0.0062	0.0486
	Std	**6.5921×10^{-05}**	0.0016	0.0058	0.0006	9.3825×10^{-04}	0.0044	0.0971
F8	Average	$-2.2572 \times 10^{+03}$	$-5.8793 \times 10^{+03}$	$-2.8442 \times 10^{+03}$	$-2.3603 \times 10^{+03}$	$-1.1147 \times 10^{+04}$	$-2.3874 \times 10^{+03}$	$-1.6823 \times 10^{+03}$
	Std	388.8521	$1.0897 \times 10^{+03}$	290.5105	**122.5179**	$1.4553 \times 10^{+03}$	459.7947	176.9365
F9	Average	0.0000	2.0687	12.9942	0.0005	2.8412×10^{-15}	14.7264	5.5881
	Std	**0.0000**	3.3984	6.5411	0.0012	1.2721×10^{-14}	6.8526	4.1836
F10	Average	2.1316×10^{-15}	1.0727×10^{-13}	0.4713	4.4445×10^{-14}	4.2643×10^{-15}	0.2989	0.3764
	Std	**1.7386×10^{-15}**	1.3980×10^{-14}	0.8689	0.0068	2.4393×10^{-15}	0.6348	0.8476
F11	Average	0.0000	0.0032	0.2189	0.0367	0.0028	0.2039	0.1543
	Std	**0.0000**	0.0068	0.1479	0.1680	0.0116	0.0881	0.0620

F12	Average	0.0020	0.0457	0.1814	0.0591	0.0021	1.0969	4.9253×10^{-07}
	Std	0.0011	0.00212	0.2564	0.0226	0.0037	1.6858	$\mathbf{1.4601 \times 10^{-06}}$
F13	Average	1.6006	0.6471	0.0031	0.2252	0.0289	0.0017	5.4868×10^{-04}
	Std	1.3405	0.2262	0.0036	0.0568	0.0251	0.0041	**0.0025**
F14	Average	2.1787	4.668	0.9980	1.5933	1.4446	1.2961	0.9880
	Std	2.4303	4.4839	$\mathbf{1.4408 \times 10^{-16}}$	0.9329	0.8192	0.5671	2.9563×10^{-16}
F15	Average	6.4854×10^{-04}	0.0044	8.2716×10^{-04}	0.0008	6.6597×10^{-04}	0.0048	0.0077
	Std	$\mathbf{4.2116 \times 10^{-05}}$	0.0092	3.2919×10^{-04}	0.0004	4.7822×10^{-04}	0.0081	0.0082
F16	Average	-1.0316	-1.0326	-1.0326	-1.0326	-1.0326	-1.0326	-1.0416
	Std	4.5168×10^{-16}	2.5459×10^{-08}	3.3831×10^{-15}	3.1413×10^{-12}	3.1413×10^{-12}	5.1535×10^{-14}	3.9452×10^{-14}
F17	Average	0.3979	0.3979	0.3978	0.3988	0.3978	0.3978	0.3989
	Std	**0.0000**	7.2878×10^{-05}	7.4515×10^{-15}	0.0012	3.2921×10^{-07}	3.3361×10^{-14}	3.7172×10^{-12}
F18	Average	3.0000	3.0000	3.0000	3.0000	3.0000	3.0000	3.0000
	Std	$\mathbf{7.1976 \times 10^{-15}}$	2.5872×10^{-05}	8.7227×10^{-14}	0.0054	9.7787×10^{-07}	1.3896×10^{-13}	3.4193×10^{-13}
F19	Average	-3.7296	-3.8617	-3.8648	-3.8748	-3.86210	-3.8638	-3.8541
	Std	0.1144	0.0021	1.0131×10^{-14}	0.0028	0.0014	$\mathbf{9.4607 \times 10^{-15}}$	0.1799
F20	Average	-2.9933	-3.2772	-3.2453	-2.9278	-3.2245	-3.2814	-3.2581
	Std	**0.0250**	0.7073	0.0586	0.2376	0.8121	0.0581	0.0621
F21	Average	-10.1532	-9.3935	-9.1356	-2.7220	-7.8594	-6.7392	-7.2580
	Std	**0.3852**	1.8508	2.0926	2.0884	2.9367	2.9772	3.3755
F22	Average	-10.4029	-10.0193	-8.8751	-3.9521	-8.7388	-8.2953	-8.9824
	Std	$\mathbf{1.3720 \times 10^{-16}}$	1.7073	3.1372	1.9058	2.6217	2.9978	2.9456
F23	Average	-10.3562	-9.7229	-9.6148	-4.6768	-9.3178	-7.3789	-8.5997
	Std	**0.3492**	2.4970	2.2967	1.2558	2.5536	3.3304	3.5350

TABLE 2.5

P-values Calculated for the Wilcoxon Rank Sum-test (Significance Level 0.05) Corresponding to the Results in Table 2.4

Ft.	AGWO	GWO	SSA	SCA	WOA	ALO	GOA
F1	N/A	6.7956×10^{-08}	6.7956×10^{-08}	6.7956×10^{-08}	0.0028	6.7956×10^{-08}	6.7956×10^{-08}
F2	N/A	6.7956×10^{-08}	6.7956×10^{-08}	6.7956×10^{-08}	6.7956×10^{-08}	6.7956×10^{-08}	6.7956×10^{-08}
F3	N/A	6.7956×10^{-08}	6.7956×10^{-08}	6.7956×10^{-08}	6.7956×10^{-08}	6.7956×10^{-08}	6.7956×10^{-08}
F4	N/A	6.7956×10^{-08}	6.7956×10^{-08}	6.7956×10^{-08}	6.7956×10^{-08}	6.7956×10^{-08}	6.7956×10^{-08}
F5	N/A	8.0065×10^{-09}	8.0065×10^{-09}	8.0065×10^{-09}	8.0065×10^{-09}	8.0065×10^{-09}	8.0065×10^{-09}
F6	N/A	8.0065×10^{-09}	8.0065×10^{-09}	8.0065×10^{-09}	8.0065×10^{-09}	8.0065×10^{-09}	8.0065×10^{-09}
F7	N/A	6.7956×10^{-08}	6.7956×10^{-08}	9.1728×10^{-08}	2.9249×10^{-05}	6.7956×10^{-08}	6.7956×10^{-08}
F8	6.7956×10^{-08}	6.7956×10^{-08}	6.7956×10^{-08}	N/A	6.7956×10^{-08}	6.7956×10^{-08}	6.5970×10^{-08}
F9	N/A	7.4517×10^{-09}	7.9043×10^{-09}	0.0096	0.3421	7.8609×10^{-09}	8.0065×10^{-09}
F10	N/A	7.6187×10^{-09}	7.9334×10^{-09}	2.2273×10^{-08}	2.3754×10^{-06}	7.9919×10^{-09}	8.0065×10^{-09}
F11	N/A	0.0402	8.0065×10^{-09}	6.6826×10^{-05}	0.3421	8.0065×10^{-09}	8.0065×10^{-09}
F12	6.7956×10^{-08}	6.7956×10^{-08}	0.0012	6.7956×10^{-08}	6.7956×10^{-08}	0.0012	N/A
F13	6.7956×10^{-08}	6.7956×10^{-08}	1.0473×10^{-06}	6.7956×10^{-08}	1.4309×10^{-07}	1.9916×10^{-04}	N/A
F14	0.0021	6.4846×10^{-05}	N/A	3.5055×10^{-07}	0.0196	0.0196	N/A
F15	0.0070	0.2503	0.0256	N/A	1.7936×10^{-04}	8.3103×10^{-04}	2.7089×10^{-04}
F16	7.9919×10^{-09}	2.4231×10^{-09}	7.9919×10^{-09}	N/A	7.9919×10^{-09}	7.9919×10^{-09}	7.9919×10^{-09}
F17	N/A	8.0065×10^{-09}	4.6827×10^{-10}	8.0065×10^{-09}	2.5497×10^{-05}	4.6827×10^{-10}	4.6827×10^{-10}
F18	7.9919×10^{-09}	0.2084	7.9919×10^{-09}	N/A	9.0065×10^{-05}	7.9919×10^{-09}	7.9919×10^{-09}
F19	N/A	2.1025×10^{-07}	4.6827×10^{-10}	8.0065×10^{-09}	8.0065×10^{-09}	4.6827×10^{-10}	0.0833
F20	0.1190	0.0052	1.7896×10^{-04}	3.9295×10^{-08}	3.4042×10^{-05}	1.0220×10^{-08}	N/A
F21	N/A	8.0065×10^{-09}	0.0198	8.0065×10^{-09}	7.9919×10^{-09}	6.0278×10^{-05}	9.1876×10^{-04}
F22	N/A	8.0065×10^{-09}	3.0335×10^{-08}	8.0065×10^{-09}	8.0065×10^{-09}	0.0792	3.3265×10^{-04}
F23	N/A	2.0446×10^{-07}	5.8842×10^{-08}	1.5747×10^{-08}	1.5006×10^{-07}	9.5900×10^{-08}	5.2113×10^{-08}

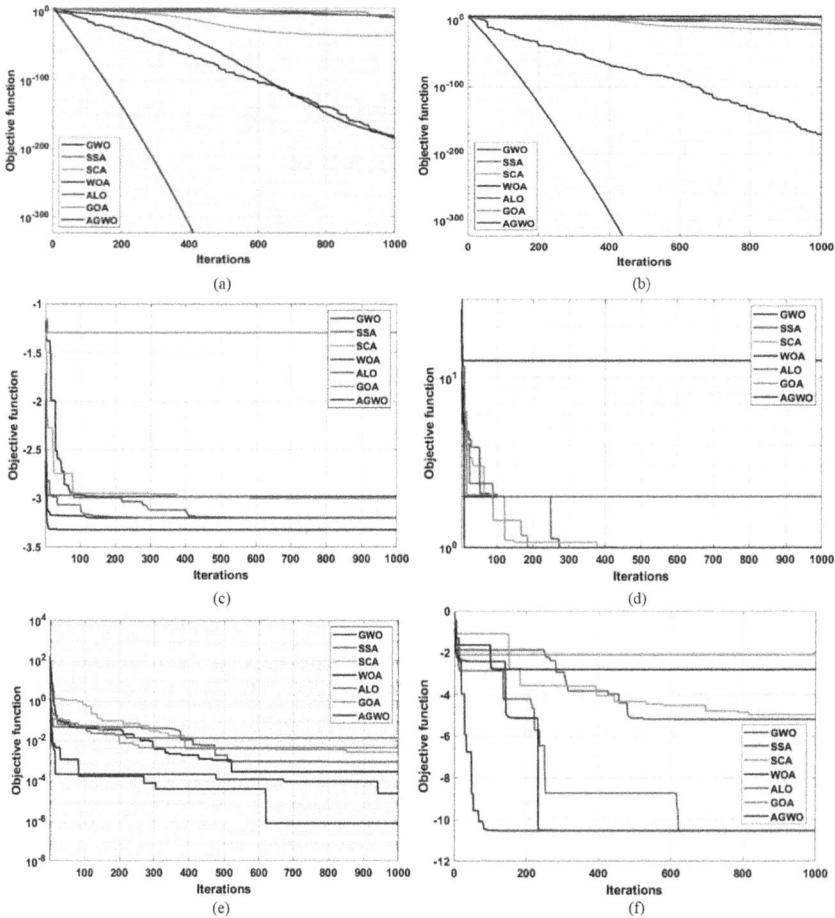

FIGURE 2.12   Comparative analysis of the proposed algorithm against existing state-of-the-art methods using benchmark functions to assess convergence.

GWO algorithm that features Gaussian mutation and position updating, for selecting parameters. The effectiveness of AGWO is evaluated against GWO, SSA, SCA, WOA, ALO, and GOA, with the convergence curves (Figure 2.12) demonstrating its quicker convergence rate on benchmark functions.

2.2.4.3   Results of CNN Model and Its Comparison with Other Classification Models

Following the methodology in Section 2.2.3, 1400 scalogram images (350 per health condition) were generated. A total of 700 images (175 per condition) were used for CNN model training, and the remaining 700 for testing (Table 2.6). Each image had dimensions 656 x 875 x 3.

TABLE 2.6
Description of the Training and Testing Datasets

S. No.	Health Condition	Training Samples	Testing Samples
1	Healthy	700 (175 x 4 =700)	700 (175 x 4 =700)
2	Splitter wear		
3	Added mass		
4	One bucket missing		

The CNN model architecture is detailed in Table 2.7. Figure 2.13(a) and 2.13(b) shows the training accuracy and loss, respectively. The model's accuracy on the test data, for each defect, is presented in Figure 2.14, showing superior performance compared to artificial neural network (ANN), support vector machine (SVM), adaptive neuro fuzzy interference system (ANFIS), and extreme learning machine (ELM) classifiers. Figure 2.15 explores the effect of varying the number of convolutional layers (five layers proved optimal).

Figure 2.16 illustrates the classification accuracy attained by the proposed AGWO algorithm in comparison to other optimization methods. Each algorithm identifies the best TVF-EMD parameters, which are subsequently utilized to create training and testing datasets for the CNN model. The findings highlight the enhanced performance of the proposed AGWO algorithm.

TABLE 2.7
CNN Architecture

S. No.	Layer Name	Layer Size
1	Input	656 x 875 x 3
2	Convolution 1	96 filters of size 11 x 11 x 3
3	Max Pooling 1	2 x 2 with stride 2
4	Convolution 2	128 filters of size 5 x 5 x 48
5	Max Pooling 2	3 x 3 with stride 2
6	Convolution 3	384 filters of size 3 x 3 x 256
7	Max Pooling 3	3 x 3 with stride 2
8	Convolution 4	192 filters of size 3 x 3 x 192
9	Max Pooling 4	3 x 3 with stride 2
10	Convolution 5	128 filters of size 3 x 3 x 192
11	Max Pooling 5	3 x 3 with stride 2
12	Fully Connected Layer	1000
13	SoftMax	–
14	Output	–

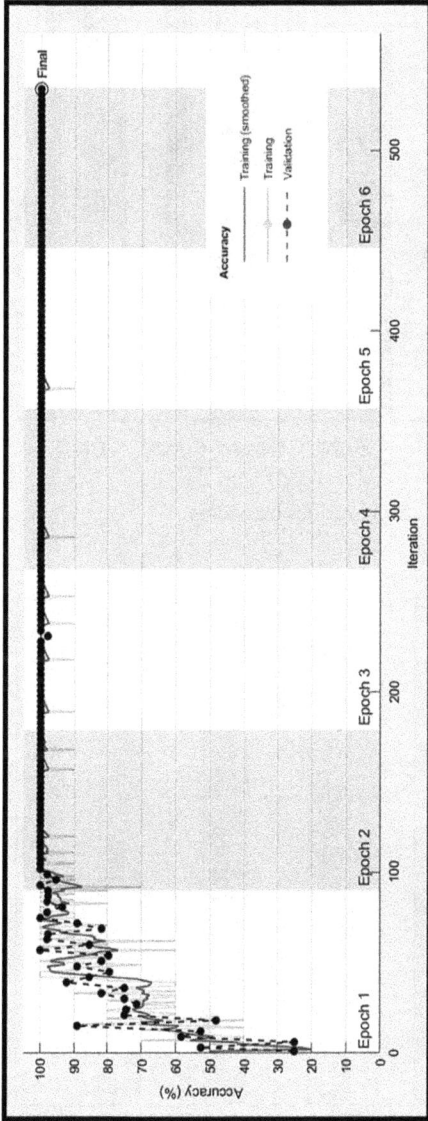
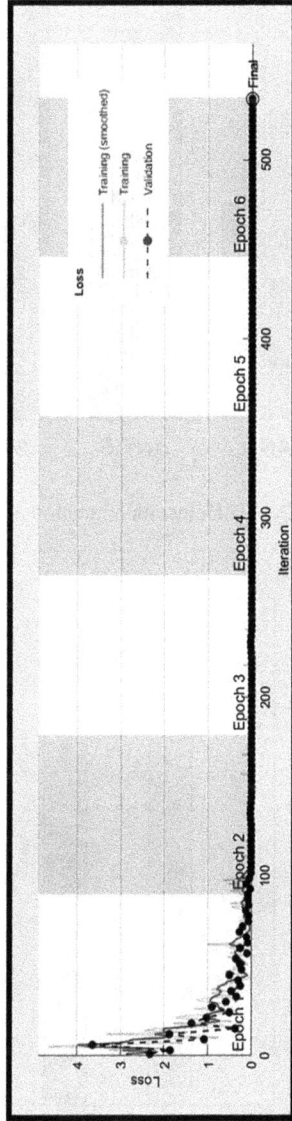

FIGURE 2.13 CNN model training performance, including (a) accuracy and (b) loss.

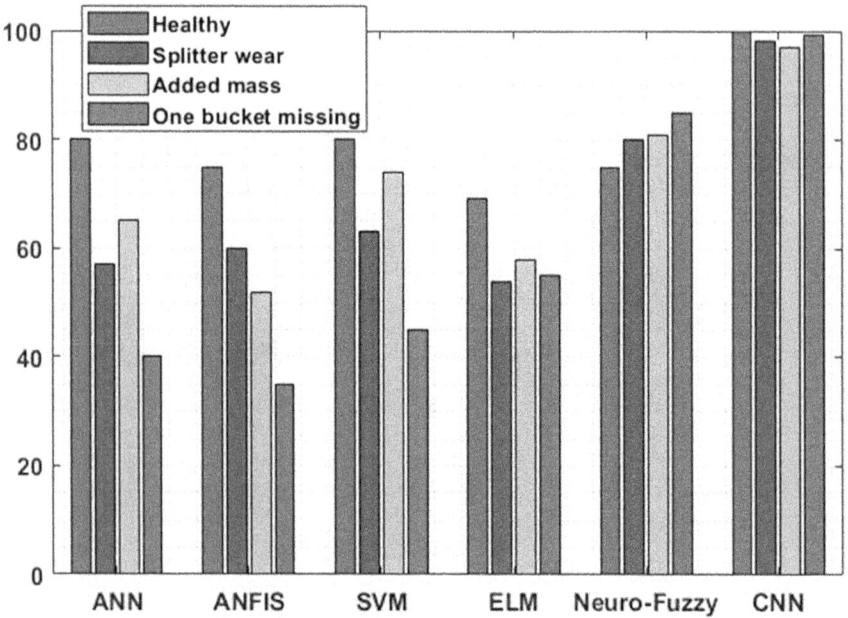

FIGURE 2.14   Defect identification accuracy using various classifiers.

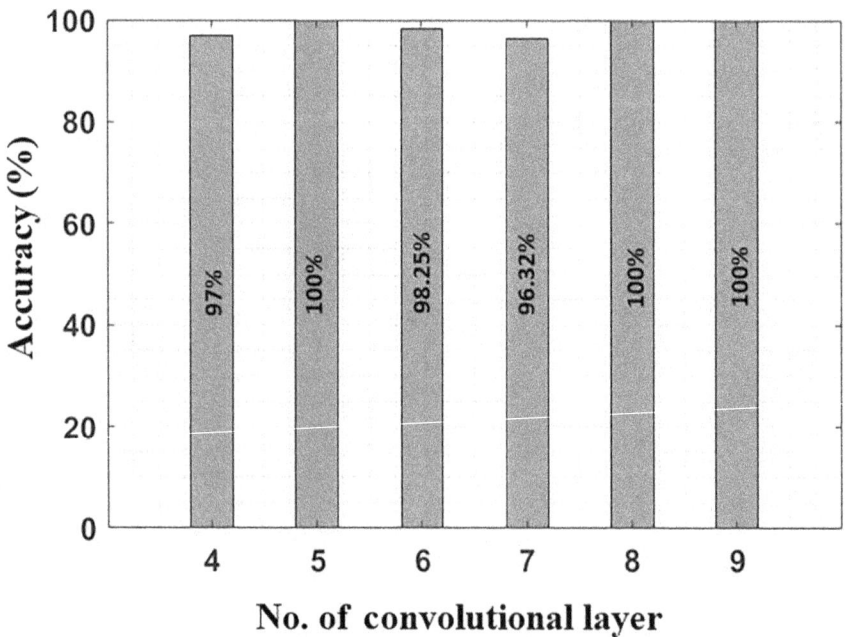

FIGURE 2.15   Accuracy achieved using different numbers of convolutional layers.

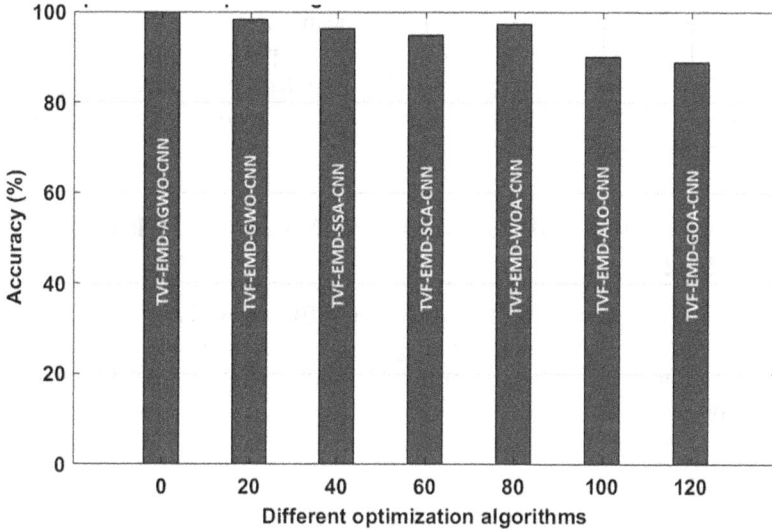

FIGURE 2.16 Accuracy in identifying defects obtained through various optimization algorithms.

2.2.5 CONCLUSION FOR CASE 1

This study uses an optimized TVF-EMD method, enhanced by a novel AGWO algorithm, for Pelton turbine bucket defect identification. AGWO makes TVF-EMD adaptive by optimally selecting ξ and n, improving the match with the input signal. The key findings are:

1. TVF-EMD's time-varying filter (a B-spline approximation filter) improves EMD performance by preserving the input signal's time-varying characteristics, crucial for effective decomposition. However, improper selection of ξ and n can negatively impact results. The proposed AGWO algorithm addresses this by adaptively selecting optimal values for these parameters.
2. This study uses the KEMI as the fitness function for TVF-EMD parameter optimization. KEMI also serves as a metric for selecting the most informative IMF, minimizing information loss.
3. AGWO's performance was evaluated against other optimization algorithms using 23 benchmark functions, comparing mean and standard deviation. AGWO achieved superior results (lowest standard deviation) on 17 functions (F1–F7, F9–F11, F15, F17–F18, and F20–F23). Wilcoxon tests confirmed this superiority as statistically significant for 15 functions (F1–F7, F9–F11, F17, F19, and F21–F23).
4. Scalograms, generated from the sensitive IMF, formed the training and test datasets for a CNN model. The model achieved 100% accuracy on the test data.

5. A comparison of classification accuracy using CNN, with different optimization algorithms, showed AGWO's superior performance. Further comparison with other learning models demonstrated CNN's superior accuracy. Analysis of CNN architecture indicated that five convolutional layers are sufficient for optimal results.

2.3 FAULT DIAGNOSIS OF BEARING IN PELTON TURBINE (CASE 2)

This section describes standard sparse filtering (SF), followed by the introduction of a novel feature extraction method: generalized normalized sparse filtering (GNSF) combined with Wasserstein distance and maximum mean discrepancy (MMD) for fault clustering. GNSF normalizes the feature matrix, while the Wasserstein-MMD approach highlights feature contributions. This methodology is applied to the Pelton turbine dataset (Chapter 4 provides details).

Vibration data was collected from the drive end bearing of a Pelton turbine (Figure 2.17) using a National Instruments data acquisition system. The turbine speed was maintained at 1100 and 1200 rpm. Five bearing conditions were studied: healthy condition (HC), inner race defect with one hole (1-IR), inner race defect with two holes (2-IR), outer race defect with one hole (1-OR), and outer race defect with two holes (2-OR) (Figure 2.18 and Table 2.8). Each condition included 400 samples at a 70 kHz sampling frequency.

2.3.1 RESULTS AND DISCUSSION

10% of the collected data was used to train the optimized sparse filter for bearing health condition diagnosis, with the remaining 90% used for validation. Input and output dimensions were set to 100, and the number of principal components ranged from 15 to 35. Figure 2.19 shows diagnostic accuracy for various values of $p(q = 2, r = 2)$. Figure 2.20 illustrates results for different q values $(p = 0.8 \text{ and } p = 3)$, indicating high accuracy at $p = 3$ ($p > q$ for optimal accuracy).

FIGURE 2.17   A pictorial view of the Pelton turbine.

1 IR   1 OR

FIGURE 2.18   Different health conditions of the Pelton turbine.

TABLE 2.8
Description of Different Health Conditions of the Pelton Turbine

S. No.	Fault Condition	No. of Samples	Condition Label
1	Healthy	400	0
2	1 seeded hole of 1 mm dia. at inner race (1 IR)	400	1
3	2 seeded holes of 1 mm dia. at inner race (2 IR)	400	2
4	1 seeded hole of 1 mm dia. at outer race (1 OR)	400	3
5	2 seeded holes of 1 mm dia. at outer race (2 OR)	400	4

Therefore, p was set to [1.8, 3.5]. Figure 2.21 supports these findings (p/q ratios of 0.5 and 1.5). Figure 2.22 shows a t-distributed stochastic neighbor embedding (t-SNE) visualization of extracted features. Table 2.9 shows the optimal normalization parameters, enabling accurate and stable bearing condition detection (Figure 2.23 shows the confusion matrix). Long short-term memory (LSTM) classifier results (accuracy and loss) are shown in Figure 2.24. These results demonstrate the effectiveness of the proposed approach for fluid machinery bearing defect diagnostics.

2.3.2   CONCLUSION FOR CASE 2

A novel unsupervised learning method for Pelton turbine bearing fault diagnosis is proposed: GNSF combined with Wasserstein distance and MMD. The method optimizes a generalized $l_{r-p/q}$ norm objective function to enhance SF regularization. Wasserstein-MMD clustering highlights feature contributions. Principal component analysis (PCA) preprocessing removes the correlation between training samples,

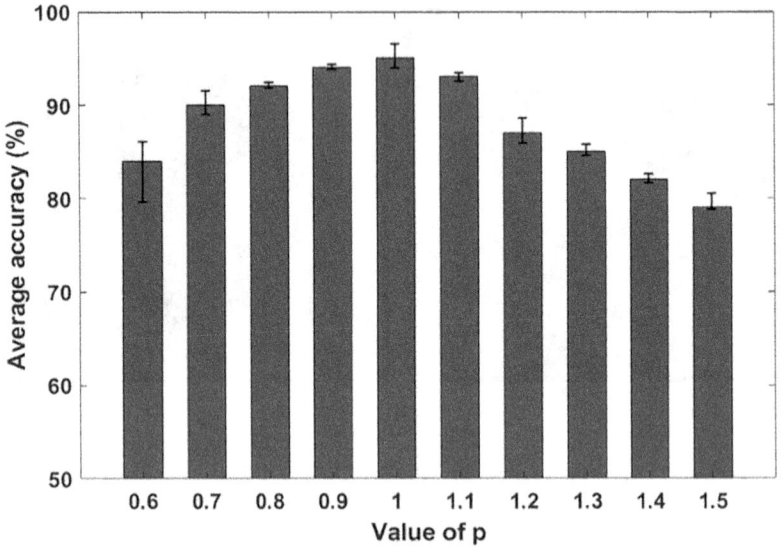

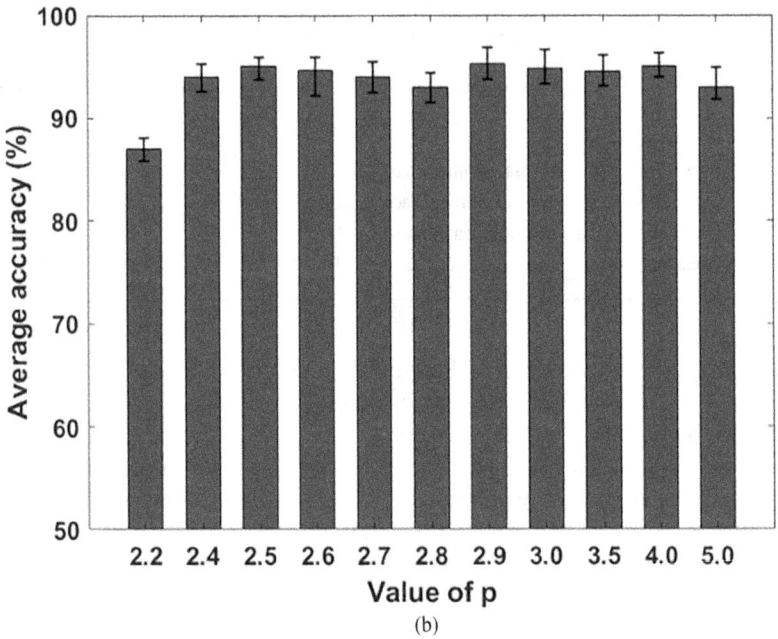

FIGURE 2.19 Diagnostic results of the Pelton turbine at various values of p with $q = 2$ and $r = 2$: (a) $p < q$ and (b) $p > q$.

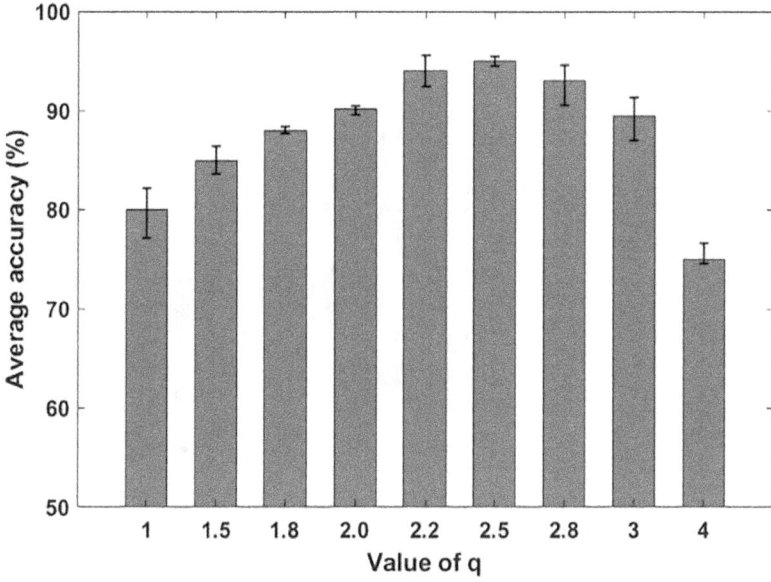

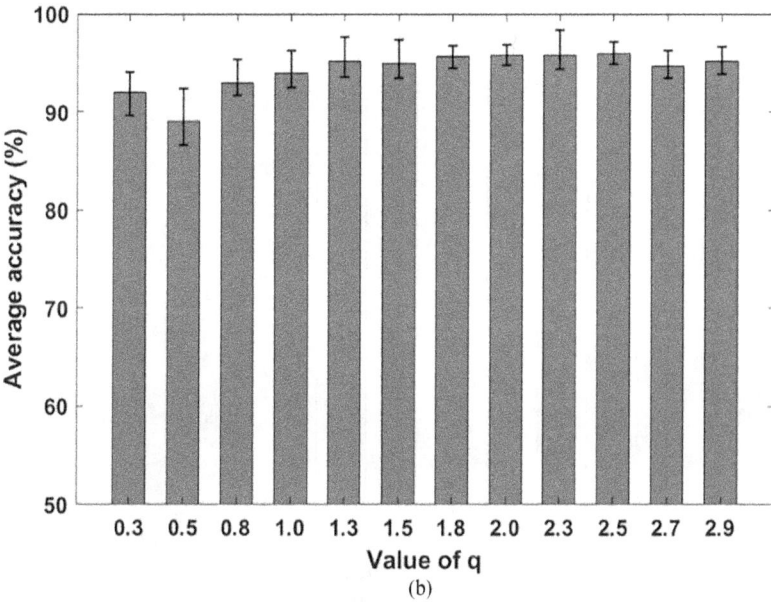

FIGURE 2.20 Diagnostic results of Pelton turbine using different normalization parameters q with $r = 2$: (a) $p = 0.8$ and (b) $p = 3$.

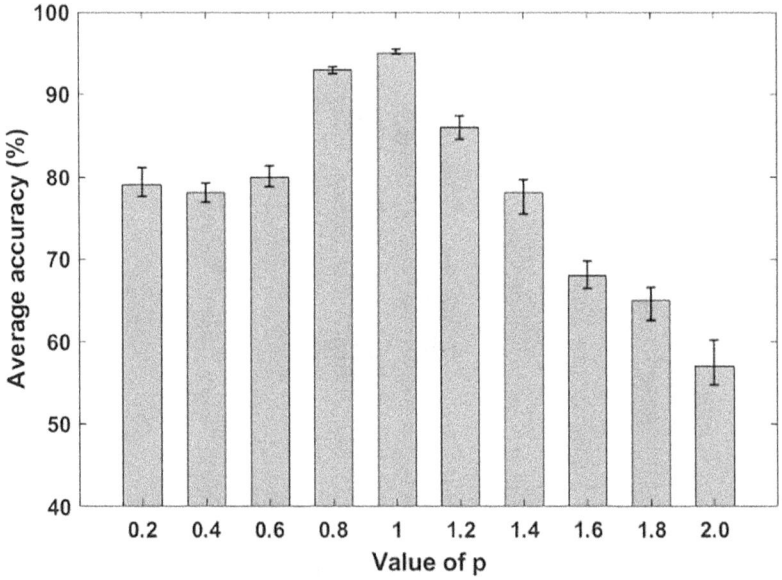

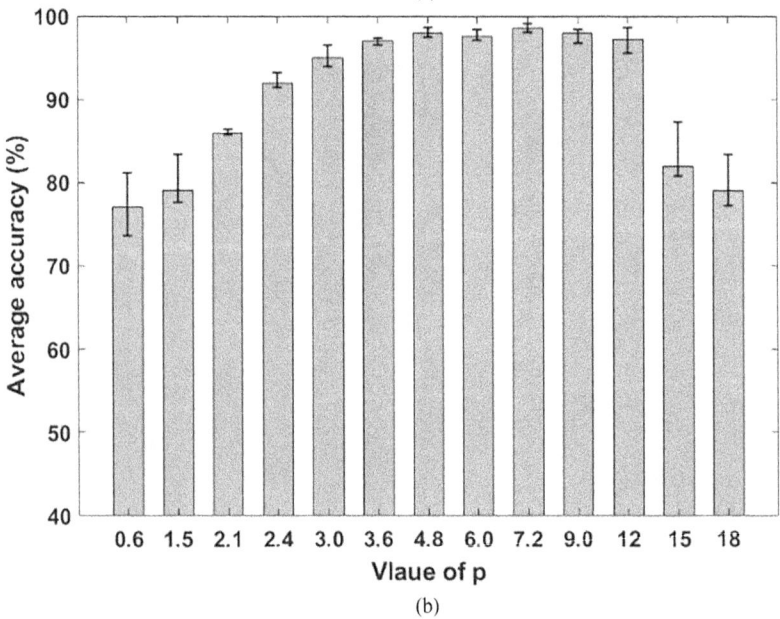

FIGURE 2.21   Diagnosis results of Pelton turbine at different values of p with $r = 2$: (a) $p/q = 0.5$ (b) $p/q = 1.5$.

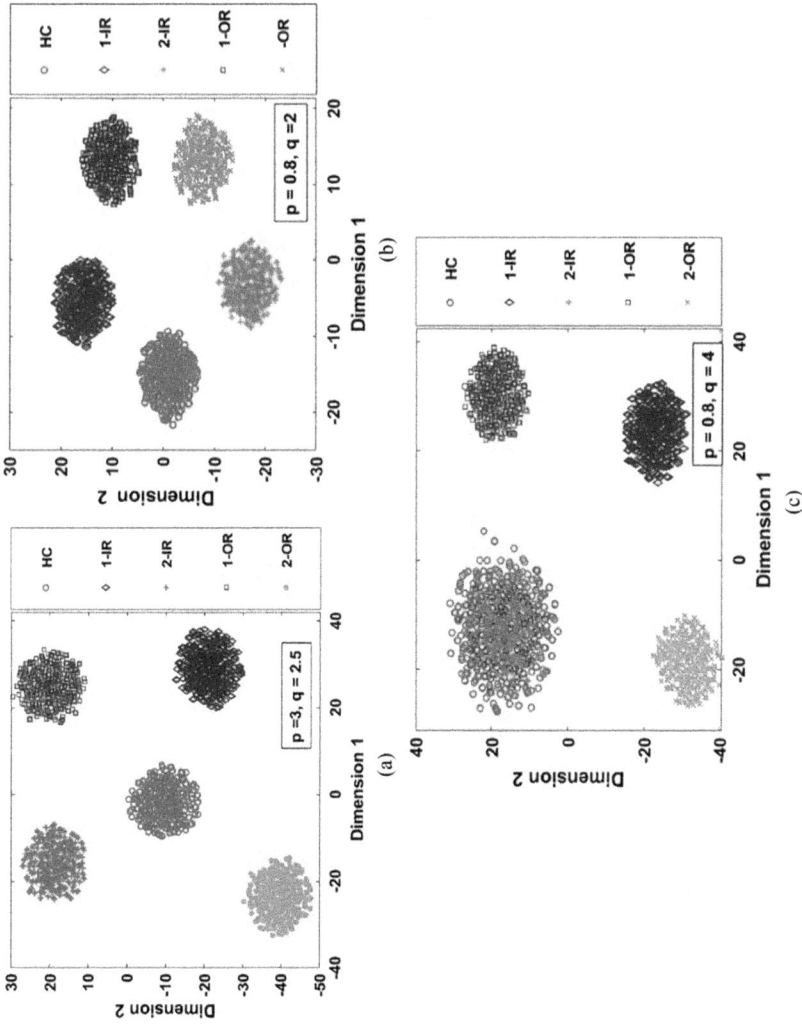

FIGURE 2.22   2D visuals of the features using t-SNE at 3 different conditions with r =2.

TABLE 2.9
Comparative Analysis of Various Sparse Filtering Methods Applied to a Pelton Turbine

Methods	No. of Training Samples	No. Health States	Computational Rime (s)	Standard Deviation (%)	Average Accuracy (%)
Standard sparse filtering	10 %	5	18.9	0.80	94.93
GNSF without PCA $(p=2.8, q=2)$	10 %	5	42.62	0.25	97.25
The proposed method $(p=2.8, q=2, 20$ PCs$)$	1 %	5	8.2	0.12	98.12
The proposed method $(p=2.8, q=2, 20$ PCs$)$	3 %	5	12.9	0.09	99.56
The proposed method $(p=2.8, q=2, 35$ PCs$)$	5 %	5	17.5	0.05	99.91

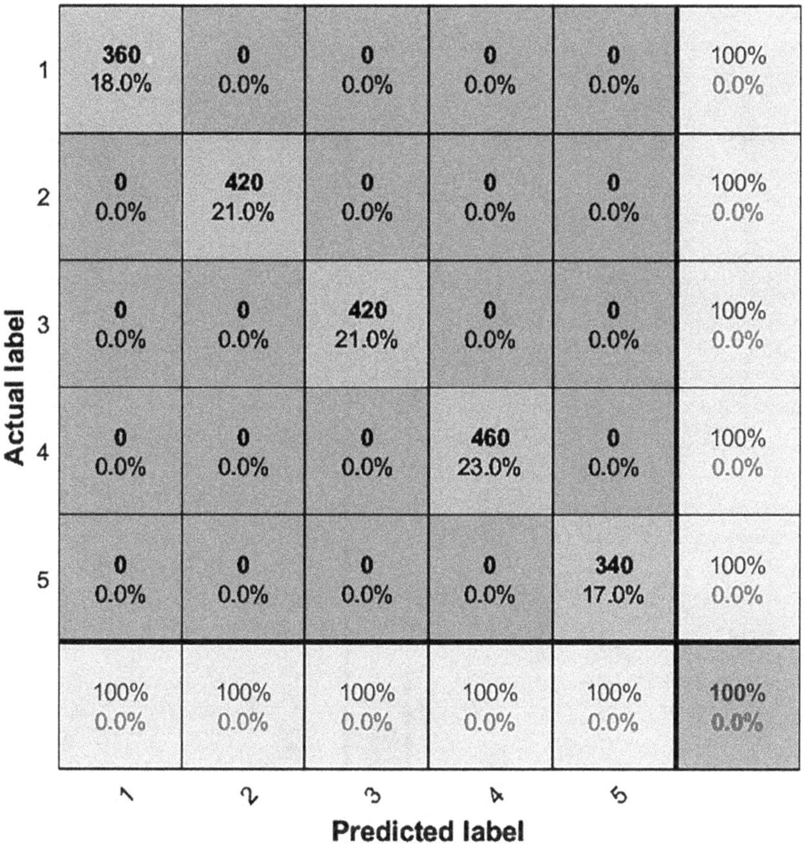

	1	2	3	4	5	
1	**360** 18.0%	**0** 0.0%	**0** 0.0%	**0** 0.0%	**0** 0.0%	100% 0.0%
2	**0** 0.0%	**420** 21.0%	**0** 0.0%	**0** 0.0%	**0** 0.0%	100% 0.0%
3	**0** 0.0%	**0** 0.0%	**420** 21.0%	**0** 0.0%	**0** 0.0%	100% 0.0%
4	**0** 0.0%	**0** 0.0%	**0** 0.0%	**460** 23.0%	**0** 0.0%	100% 0.0%
5	**0** 0.0%	**0** 0.0%	**0** 0.0%	**0** 0.0%	**340** 17.0%	100% 0.0%
	100% 0.0%	100% 0.0%	100% 0.0%	100% 0.0%	100% 0.0%	**100%** **0.0%**

Actual label

1   2   3   4   5

Predicted label

FIGURE 2.23   Confusion matrix.

and an LSTM classifier identifies bearing faults. Pelton turbine data validates the method's robustness, leading to the following conclusions:

1. Optimized SF parameters ensure accurate and reliable results by adaptively extracting relevant features from the vibration signal.
2. Wasserstein distance with MMD is used for feature clustering, highlighting each feature's contribution. Comparison with traditional methods demonstrates the superiority of this novel approach.
3. The proposed method effectively identifies centrifugal pump health conditions even with limited training data; for example, achieving 99.91% accuracy with only 5% of the samples used for training.
4. The proposed method uses GNSF to extract discriminative features from Pelton wheel vibration data, which are then clustered using Wasserstein distance with MMD to facilitate accurate fault classification.
5. GNSF offers a wider range of normalization parameters, resulting in more accurate and robust performance compared to traditional SF.

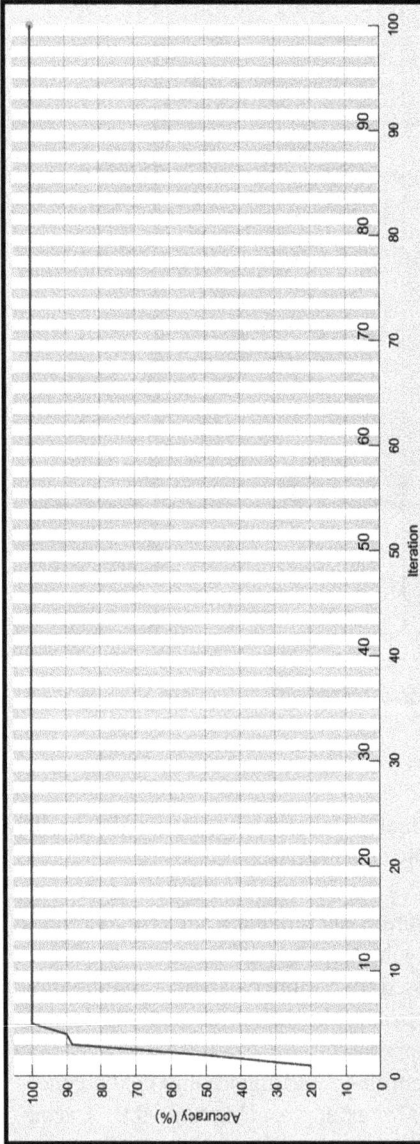

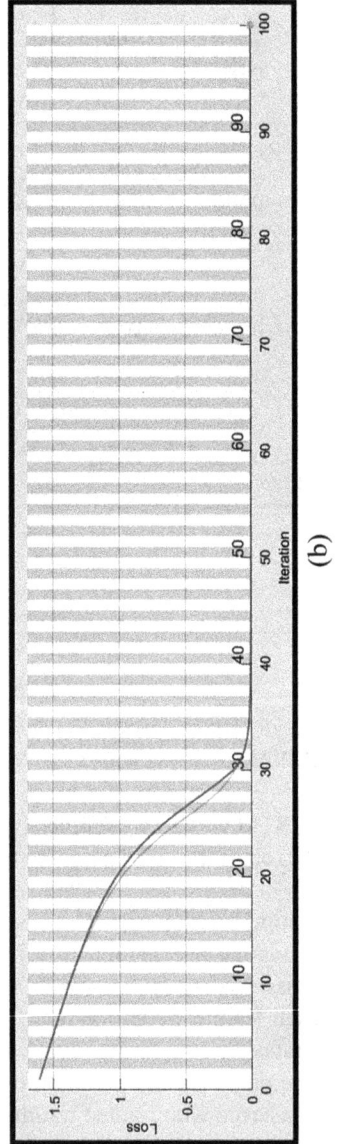

FIGURE 2.24 Training performance of LSTM classifier for Pelton turbine (a) accuracy and (b) loss.

TABLE 2.10
Comparison of Fault Schemes of Cases 1 and 2

Methods	Bucket Defects		Bearing Defects	
	Average Accuracy (%)	Computational Time (s)	Average Accuracy (%)	Computational Time (s)
TVF-EMD-AWGO-CNN	100	25.84	100	27.39
GNSF (p = 2.8, q = 2, 35 PCs)	98.52	18.67	99.91	17.5

2.4 COMPARISON OF METHODOLOGIES PROPOSED IN CASES 1 AND 2

While Case 1's fault diagnosis scheme (for Pelton turbine bucket defects) demonstrates robustness and superior performance, it requires manual feature extraction. In contrast, Case 2's scheme (for bearing defects) is fully automated. To compare these schemes fairly, Case 1's method was applied to bearing defect data, and Case 2's method to bucket defect data. Table 2.10 presents the resulting accuracy and computational times.

Table 2.10 shows that method in Case 1 achieves slightly higher accuracy, while method in Case 2 has a shorter computation time.

2.5 SUMMARY

This chapter investigates two Pelton turbine defect cases: bucket and bearing defects. For bucket defects, an AGWO algorithm, using KEMI as the fitness function, optimizes TVF-EMD parameters. The optimal solution (minimum KEMI) is used to generate scalograms for training and testing a CNN classifier. For bearing defects, a GNSF method, incorporating Wasserstein distance and MMD for fault clustering, is proposed, normalizing the feature matrix to highlight feature contributions.

3 Fault Diagnosis of the Francis Turbine

3.1 INTRODUCTION

The Francis turbine is an inward-flow reaction turbine that can achieve efficiencies greater than 95% [63]. It consists of a spiral casing, guide vanes, stay vanes, runner blades, and a draft tube. Water flows into the spiral casing, moving through stay vanes for smoothing and adjustable guide vanes that manage the angle of attack on the runner blades. The runner blades have two sections: a lower bucket-like section (impulse action) and an upper section (reaction force). The runner thus utilizes both pressure and kinetic energy. The draft tube recovers pressure energy before the tailrace, compensating for the low-energy state of the water exiting the runner blades [64–66].

Increased power demands on Francis turbines lead to higher hydraulic forces and stresses on components (runners, spiral casing, vanes), potentially causing fatigue, misalignment, bearing defects, and cracks [67–69]. Bearing defects are a major cause of turbine failure, characterized by quasiperiodic and periodic impulses [67–69]. Acoustic signal analysis is a useful contactless method for defect detection, particularly beneficial due to the hazardous operating conditions and potential for leakage that make accelerometer-based measurements difficult [70–73]. However, environmental noise and signal degradation can obscure fault features. Blind deconvolution, such as minimal entropy deconvolution (MED) [74], can enhance weak fault characteristics. MED, initially used in seismic signal processing [74], increases kurtosis for weak impulses while reducing it for noise [75]. While MED improves spectral kurtosis (SK) for fault detection [76–78], excessive kurtosis can create spurious impulses. Optimal filter length is also crucial. Improvements to MED include replacing kurtosis with skewness [79, 80], correlation kurtosis maximum correlated kurtosis deconvolution (MCKD) [81], optimal minimum entropy deconvolution adjusted (OMEDA) and multipoint OMEDA (MOMEDA) (MOMEDA) [82], minimum entropy deconvolution based on impulse norm (MIND) [83], and autocorrelation impulse harmonic to noise (AIHN) [84], or optimizing filter coefficients using algorithms such as particle swarm optimization (PSO) [85] and l_o norm optimization [86]. These enhanced methods have demonstrated improved rotating machinery diagnostics.

Research indicates that improper filter length in MED can transform periodic impulses into single pulses, with longer lengths generally leading to higher kurtosis values [85–88]. However, a definitive method for filter length selection remains elusive with empirical formulas yielding inconsistent results. Li et al. [89] introduced an adaptive MED that employs a modified power spectrum kurtosis (MPSK)

DOI: 10.1201/9781003614821-3

index for the diagnosis of wind turbines; however, the outcomes were not ideal, and the impact of filter length was not thoroughly investigated. Some studies rely on empirical filter length selection without considering its impact on diagnostic accuracy [90–92].

This chapter presents a novel automated method for determining optimal MED filter length to improve rotating machinery diagnostics, addressing limitations in existing approaches. The autocorrelation function, used as an energy measurement index to identify fault-related periodic impulses, guides the filter length selection. While various optimization techniques exist [93–95], this study employs the Aquila optimizer (AO) [96] to find the optimal filter length, maximizing the autocorrelation function. Experiments on a Francis turbine bearing demonstrate the method's effectiveness in enhancing weak fault features in acoustic signals.

3.2 THEORETICAL BACKGROUND

3.2.1 MINIMAL ENTROPY DECONVOLUTION (MED)

MED seeks to develop an inverse filter based on a typical signal transfer function [75]. Eq. (3.1) illustrates the vibration signal, x_N, obtained from rotating machinery.

$$x_N = (p + m) * h \qquad (3.1)$$

where p stands for periodic fault impulses, m signifies noise, h indicates a parameter that influences harmonics and transmission, and * represents convolution. The process of the MED filter is illustrated in Figure 3.1.

The MED method uses a low-entropy sparse pulse sequence, s, as input. The MED system increases the entropy of the resulting signal, x. Deconvolution seeks to identify the finite impulse response (FIR) filter, f (length L), such that the filtered output, y, approximates the original input, s, as shown in Eq. (3.2).

$$y(j) = \sum_{l=1}^{L} f(l) x(j-l) \approx s(j) \, j = 1,2,\ldots,N \qquad (3.2)$$

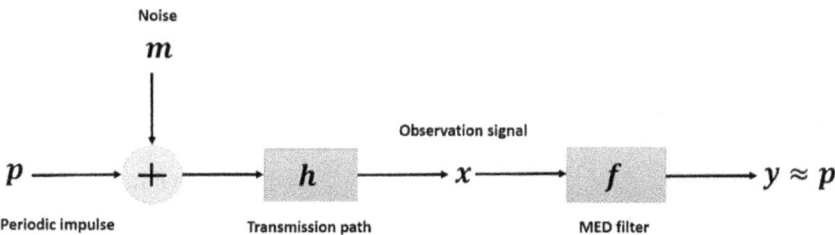

Noise

m

Observation signal

$p \longrightarrow + \longrightarrow h \longrightarrow x \longrightarrow f \longrightarrow y \approx p$

Periodic impulse   Transmission path   MED filter

FIGURE 3.1 Working of MED filter.

MED is typically implemented using either an eigenvector method or an objective function method. This research utilizes the objective function method, implemented via kurtosis calculation, as detailed in Eq. (3.3).

$$O_4\left[f(l)\right] = \sum_{j=1}^{N} y^4(j) \left/ \left[\sum_{j=1}^{N} y^2(j)\right]^2\right. \tag{3.3}$$

To determine the optimal filter, we calculate the first derivative of Eq. (3.3) and set it equal to zero. In this context, N represents the total length of the data set under consideration.

$$\partial O_4\left[f(l)\right] / \partial f(l) = 0 \tag{3.4}$$

Eq. (3.2) can also be represented in the form of matrix as

$$y = X_0^T f \tag{3.5}$$

$$\text{where } X_0 = \begin{bmatrix} x_1 & x_2 & x_3 \cdots \cdots & x_N \\ 0 & x_1 & x_2 & \cdots & \cdots & x_{N-1} \\ 0 & 0 & x_1 & \cdots & \cdots & x_{N-2} \\ \vdots & \vdots & \vdots & \ddots & \cdots & \vdots \\ 0 & 0 & 0 & \cdots & \cdots & x_{N-L+1} \end{bmatrix}_{L \times N}$$

Substituting Eq. (3.3) and Eq. (3.4) into Eq. (3.5) and by further simplification, Eq. (3.6) is obtained.

$$f = \frac{\sum_{j=1}^{N} y_j^2}{\sum_{j=1}^{N} y_j^4} \left(X_0 X_0^T\right)^{-1} X_0 \left[y_1^3 y_2^3 \dots y_N^3\right]^T \tag{3.6}$$

The procedure followed by the MED filter is as follows:

Step 1: Initializing $f^0 = (0,1,0,\dots,0)^T$ and also feed the raw vibration signal x in order to obtain X_0^T.

Step 2: Set the values of filter parameters such as filter length (L), maximum iteration $(m_{\max})$ and convergence error (ξ).

Step 3: Using Eq. (3.5), compute $y^m(j)$ by substituting X_0^T and filter coefficient $f^m(l)$ in order to get f^{m+1} through Eq. (5.6).

Step 4: Compute the error $\Delta E = \left| O_4\left(f^{m+1}\right) - O_4\left(f^m\right) \right|$ utilizing Eq. (3.3).

Step 5: When $m < m_{max}$ and $\Delta E < \xi$ the iteration using Eq. (3.3) will be continued otherwise save the final output obtained from Eq. (3.5) and Eq. (3.6).

3.2.2 INFLUENCE OF FILTER LENGTH ON OUTPUT RESULTS

The three basic components of the signal are represented in Eq. (3.7).

$$x(t) = u(t) + n(t) + h(t) \tag{3.7}$$

Figure 3.2(a) illustrates the fault impulse $u(t)$, which has a fault sampling interval of 30 points. As depicted in Figure 3.2(b), noise $n(t)$ is introduced into the fault signal, with an impulse to noise energy ratio of 0.23. The resulting signal x(t) incorporates a harmonic component $h(t) = 0.1\sin\left(2\pi f_1 t\right) + 0.2\sin\left(2\pi f_2 t\right) + \sin\left(2\pi f_3 t\right)$, where $f_1 = 4f_2 = 2f_3 = \frac{1}{15}$. The final signal comprises 2000 sample points or data points.

Figure 3.3 demonstrates the application of MED to the resultant signal using filter lengths of $L = 130$ and $L = 131$. When $L = 130$, the filtration produces a

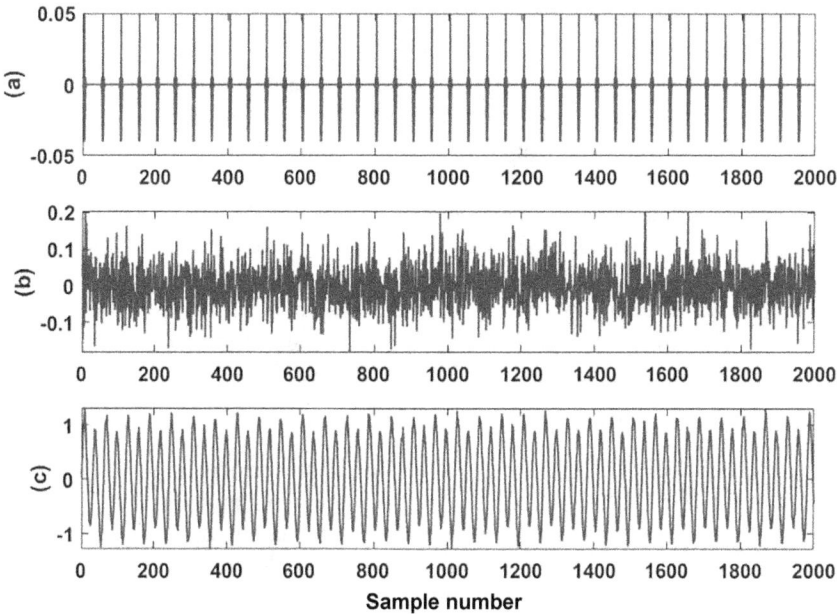

FIGURE 3.2 Simulated signal having fault (a) fault impulse signal $u(t)$, (b) fault impulse signal with noise $n(t)$, and (c) resultant signal of $u(t), n(t)$, and $h(t)$.

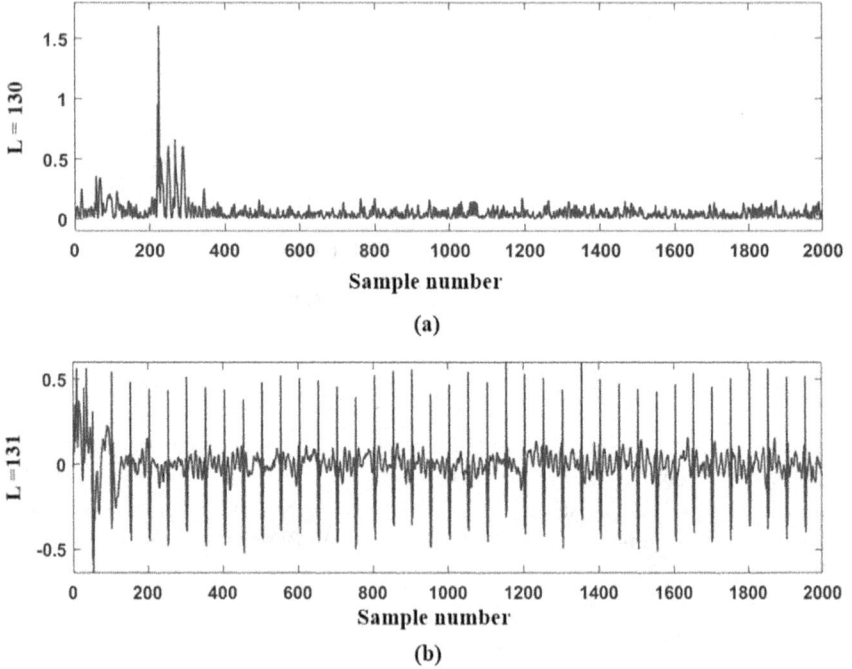

FIGURE 3.3   The filtered signals at filter length (a) $L = 130$ and (b) $L = 131$.

single impulse, which is not ideal for diagnosing various faults in rotating machinery. Ideally, a series of periodic pulses representing fault characteristics should be obtained. These periodic pulses are clearly visible when $L = 131$. It's important to note that changing the filter length by just one unit results in drastically different outputs.

These observations underscore the significance of filter length in MED, as it directly influences the output. Consequently, selecting the appropriate filter length is a crucial step in the MED process. To resolve the issue of single-pulse extraction, it's essential to conduct a thorough investigation of the filter length determination process, as relying on experience alone may lead to inaccurate results.

3.2.3   Optimum Filter Length Selection

To improve the MED filter's capacity to amplify faint periodic impulses, it is necessary to create a fitness function that assesses the periodicity of the filtered signal. This research utilizes autocorrelation energy to identify the best filter length.

3.2.3.1   Autocorrelation Analysis

Autocorrelation analysis serves as a useful tool for illustrating how a signal correlates with itself over various time intervals, which makes it highly effective for identifying periodic patterns in the signal. Eq. (3.8) illustrates the observed signal, wherein $s(t)$ indicates a sinusoidal periodic signal, and $n(t)$ stands for white Gaussian noise.

$$x(t) = s(t) + n(t) = A\sin(\omega_0 t + \varnothing) + n(t) \tag{3.8}$$

After $x(t)$ has been passed through the autocorrelation analysis, it yields:

$$R_x(\tau) = E\left[x(t)x(t-\tau)\right] = R_s(\tau) + R_n(\tau) + R_{sn}(\tau) + R_{ns}(\tau) \tag{3.9}$$

Here, $s(t)$ and $n(t)$ are not dependent on each other that is why $R_{sn}(\tau) = R_{ns}(\tau) = 0$, which results into:

$$R_x(\tau) = R_s(\tau) + R_n(\tau) = \lim_{T\to\infty} \frac{1}{2T} \int_{-T}^{T} \left[s(t)s(t-\tau)\right] dt + R_n(\tau)$$

$$= \frac{A^2}{2} \cos(\omega_0\tau) + R_n(\tau) \tag{3.10}$$

Here, A represents amplitude, ω_0 denotes angular frequency, and $\varnothing$ signifies the initial phase. The noise is represented by $n(t)$, and $R_n(t)$ is concentrated around $\tau = 0$, as illustrated in Figure 3.4. According to Eq. (5.9), the signal in $R_s(\tau)$ shares the same angular frequency ω_0 as $s(t)$. As τ increases, $R_x(\tau)$ primarily reflects $R_s(\tau)$, allowing $R_x(\tau)$ to be used for determining the amplitude and frequency of $s(t)$, as shown in Figure 3.5.

3.2.3.2 Development of Fitness Function

A key limitation of MED is its use of a fixed filter length for continuous computation, leading to results dependent on filter length. The fitness function's main role

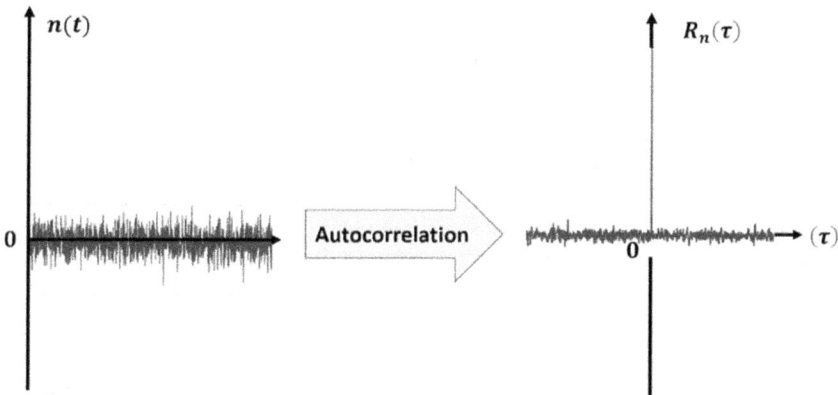

FIGURE 3.4 Autocorrelation analysis of noisy signal.

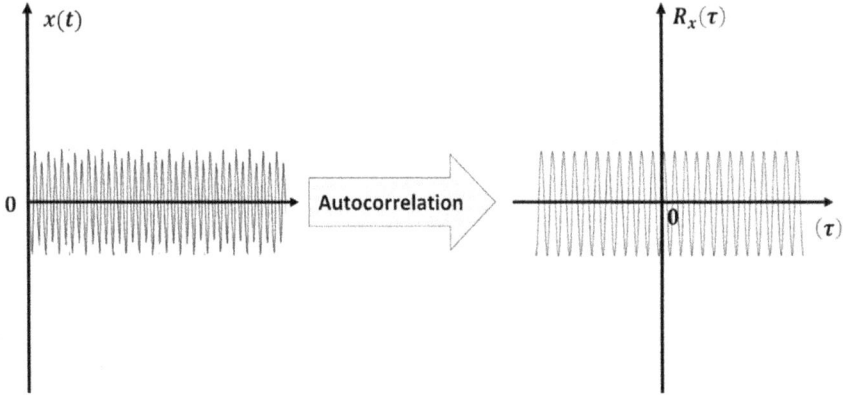

FIGURE 3.5 Autocorrelation analysis of observed signal.

is to provide an index quantifying the filtered signal's periodicity across different filter lengths.

For a specific filter length L, let y_L represent the output signal. The remaining components of the raw signal can be expressed as:

$$s_L = x_N - y_L \qquad (3.11)$$

Here, x_N denotes the unprocessed signal. The fitness function is formulated according to Eq. (3.12).

$$L_\mu = \frac{\displaystyle\sum_{n=1}^{N} R_{y_L}^2 (n)}{\displaystyle\sum_{n=1}^{N} R_{S_L}^2 (n)} \qquad (3.12)$$

Here, N represents the data length, and $R_{y_L}(.)$ and $R_{S_L}(.)$ represent the output and residual autocorrelation functions, respectively. Periodic impulses in the output signal result in periodic autocorrelation. In these situations, MED improves the kurtosis of weak impulses while reducing harmonics and noise from the system, thereby elevating the energy ratio between the filtered signal and the residual signal.

Conversely, a single-pulse output signal results in a residual signal dominated by noise, except for the main impact. This concentrates autocorrelation amplitude near time zero, leading to a lower energy ratio. The value of L_μ determines if the output constitutes a periodic pulse. To prevent interference, the autocorrelation value at time zero needs to be omitted.

The effectiveness of L_μ has been demonstrated using a simulated signal represented in Eq. (3.7). Figure 3.6 illustrates the computed kurtosis and L_μ of the output signal at filter lengths ranging from $L = 2$ to $L = 500$. Figure 3.6(a) clearly shows that kurtosis increases with filter length, consistent with results presented

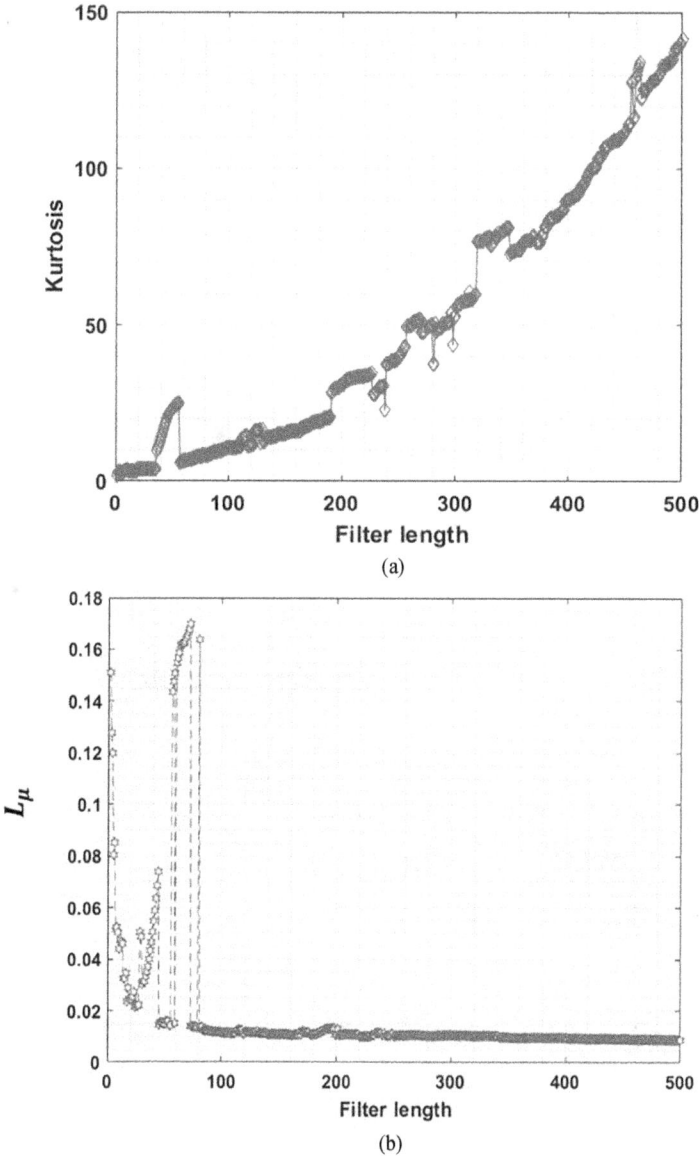

(a)

(b)

FIGURE 3.6 (a) Kurtosis and (b) fitness function at various filter lengths.

in Ref. [87]. This behavior arises because the MED fitness function maximizes kurtosis, a strong indicator of impact characteristics. Nonetheless, ongoing impulses in the output signal do not have a direct relationship with changes in kurtosis; elevated kurtosis can amplify an individual pulse within the total signal.

Figure 3.6(b) shows the variation of L_μ at different filter lengths, initially increasing, then decreasing, and finally stabilizing. As demonstrated in Figure 3.3, there's a significant difference in output between filter lengths $L = 130$ and $L = 131$. Figure 3.7 presents outputs at filter lengths $L = 100, 185,$ and 186.

The results clearly indicate that a larger L_μ, the MED filter performs better, obtaining strong periodic impact characteristics. However, beyond a certain filter length (e.g., $L = 186$), L_μ decreases continuously, and the output signal exhibits a single impulse. At this point, MED becomes insufficient for enhancing periodic weak signals. Thus, L_μ can serve as a fitness function for selecting the optimal filter length for the MED filter.

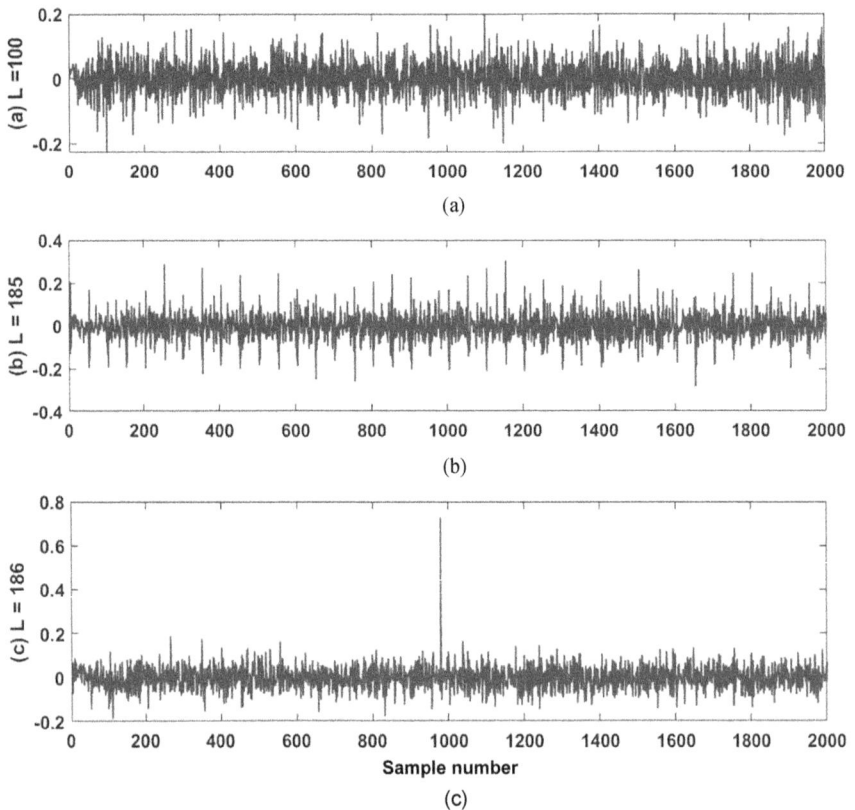

FIGURE 3.7   Signals at filter lengths of (a) L = 100, (b) L = 185, and (c) L = 186.

3.2.3.3   Aquila Optimizer (AO)

The AO algorithm [96] mimics the hunting tactics of the Aquila bird, which includes processes such as initialization, evaluating the fitness function, and updating solutions. This research applies AO to enhance the L_μ fitness function for MED, determining the ideal integer filter length that is shorter than the raw signal length. AO's search strategy involves four approaches: expanded and narrowed exploration and expanded and narrowed exploitation, aimed at finding near-optimal and best solutions based on the fitness function. It iteratively adjusts solution positions until a specified termination criterion is met.

Figure 3.8 illustrates the effect of population size (20, 40, 60, 80, and 100) on the results, analyzing 100 iterations for each size.

Figure 3.8 shows that smaller population sizes yield suboptimal fitness function (L_μ) values. Increasing the population size increases L_μ, reaching a maximum of 0.4867 at L = 159 (similar to Figure 3.6(b)). The relationship between population size and L_μ is logarithmic, suggesting saturation beyond a certain population size. While smaller populations may not maximize L_μ, some periodic impulses might still be detected due to filtering. Since AO consistently seeks an "optimal solution," even with low L_μ values (resulting in single-impulse outputs), a population size of 100 and 100 iterations were used (Figure 3.9 details the optimal filter length selection).

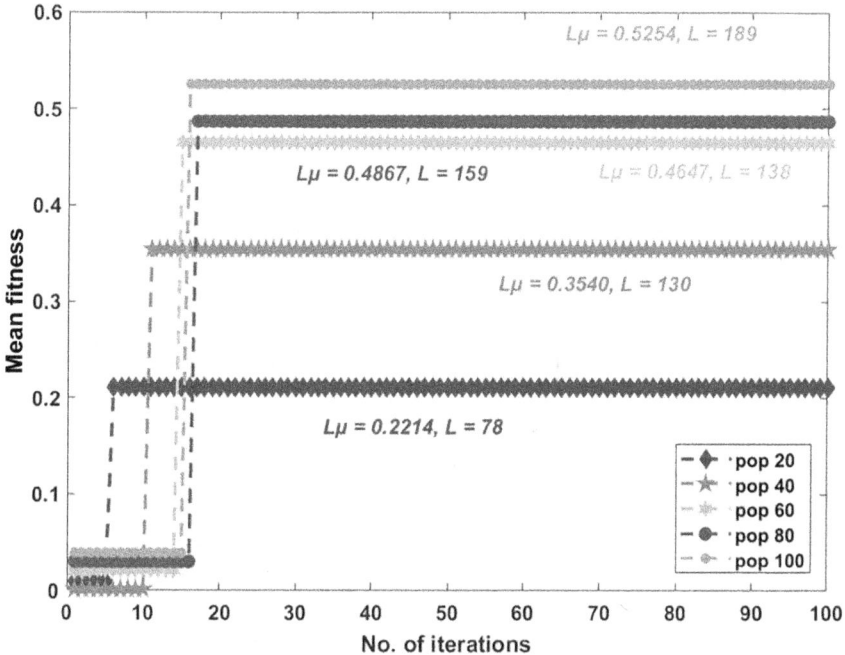

FIGURE 3.8   Optimal filter length for different populations at 100 iterations.

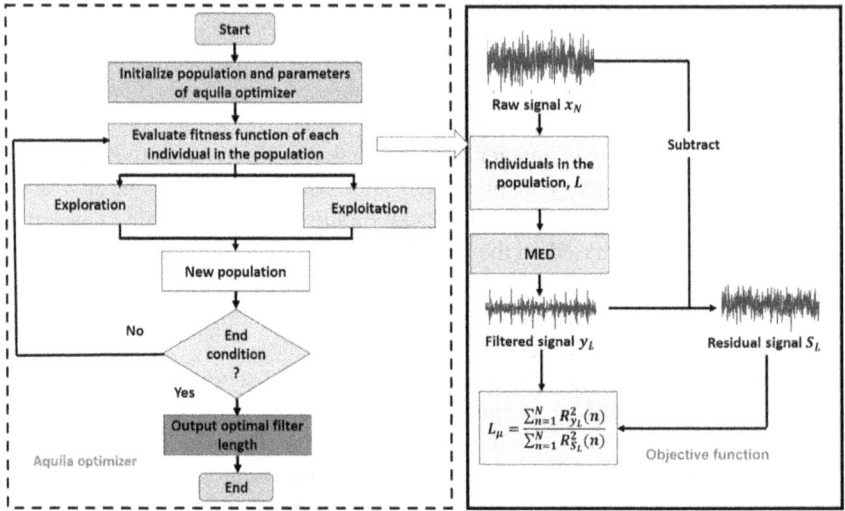

FIGURE 3.9   Procedure for obtaining the optimized filter length.

The basic steps are as follows:

1. The AO initialization step sets AO parameters and generates a random population of potential solutions, with encoding length determined by the raw signal length.
2. The L_μ metric assesses the quality of each solution within the AO population. The exploration and exploitation strategies of AO subsequently uncover promising near-optimal and top solutions.
3. Following N iterations, the algorithm concludes and provides the best solution identified, which indicates the optimal length of the filter.

The flow chart of the proposed scheme for fault identification is shown in Figure 3.10.

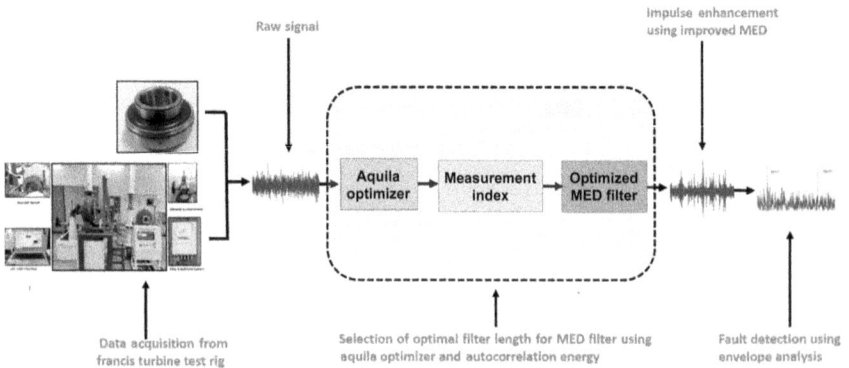

FIGURE 3.10   Flowchart of the proposed fault detection method.

3.3 FAULT DIAGNOSIS OF FRANCIS TURBINE

The proposed method was applied to data from a Francis turbine test rig (Figure 3.11). The turbine operated at 2350 rpm and 3040 rpm (specifications are shown in Table 3.1).

The Francis turbine (refer to Figure 3.11) is made up of a spiral casing, guide vanes, stay vanes, runner blades, and a draft tube. The guide and runner blades are attached to a shaft that uses UC 206 bearings. Water flows through the spiral casing, guided by stay and guide vanes (the latter controlling power output by adjusting the angle of attack on the runner blades). Guide vane water impact can cause vibrations, leading to potential issues, such as shaft misalignment, runner or vane damage, and bearing defects, which are the focus of this study. Five bearing conditions were analyzed: healthy and seeded defects (1 mm diameter spalls created via wire electrical discharge machining (WEDM) of one or two spalls on either the inner (1-IR, 2-IR) or the outer (1-OR, 2-OR) race (Figure 3.12). UC 206 bearing specifications are detailed in Table 3.2.

Acoustic data from the faulty Francis turbine bearing was recorded using a microphone near the bearing, employing a National Instruments 24-bit, four-channel data acquisition system at a 70 kHz sampling rate. Subsequent analysis used 0.1-second segments (7000 data points).

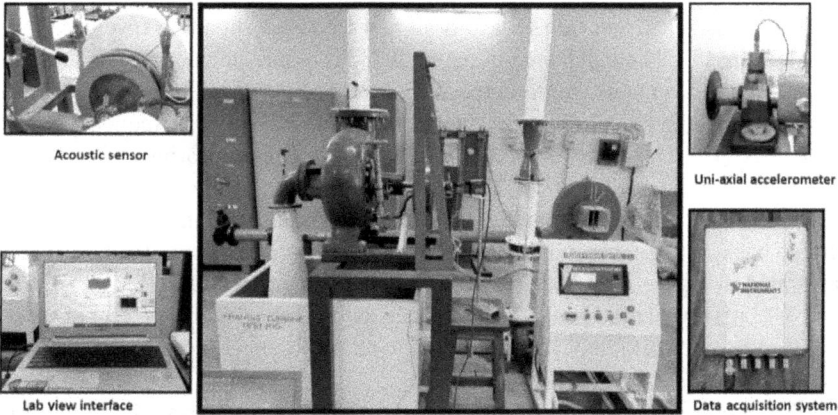

FIGURE 3.11 Test rig of Francis turbine.

TABLE 3.1
Specification of Francis Turbine

Maximum output	3 kW
Maximum discharge	400 l/min

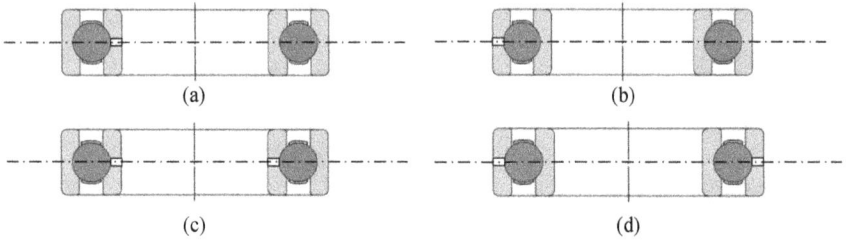

(a)   (b)

(c)   (d)

FIGURE 3.12 Schematic of defective bearings: (a) one spall defect on the inner race (1-IR), (b) one spall defect on the outer race (1-OR), (c) two spall defects on the inner race (2-IR), and (d) two spall defects on the outer race (2-OR).

TABLE 3.2
Specification of UC 206 Bearing

Type	Inner Diameter	Outer Diameter	No. of Balls	Contact Angle
UC-206	30 mm	62 mm	9	0°

The inner race defect signal waveforms and their corresponding envelope spectra (1-IR and 2-IR) are depicted in Figure 3.13. The time-domain signals for both acoustic measurements, represented in Figure 3.13 (a) and 3.13 (c), exhibit a series of periodic pulses. In the envelope spectrum, the ball pass frequency of the inner race (BPFI) is observed at 397 Hz, with harmonics appearing at 795 Hz for the 1-IR scenario, and at 397 Hz and 2 BPFI (795 Hz) for the 2-IR scenario, as illustrated in Figure 3.13 (b) and 3.13 (d). These impulses are masked by considerable background noise, showing a gradual decrease in their amplitude, as demonstrated in Figure 3.13 (a) and 3.13 (c). The envelope spectra in Figure 3.13 (b) and 3.13 (d) reveal that the BPFIs are hidden among other frequencies, complicating the direct extraction of fault information related to inner race defects from the acoustic signals.

The proposed method was used to detect inner race-bearing defects in the Francis turbine. AO optimization determined an optimal filter length of $L = 100$ for single inner race spall defects (1-IR), yielding $L_\mu = 0.8187$ (Figure 3.14a). Figure 3.14b shows L_μ values for various filter lengths, mostly below 1. For two inner race spalls (2-IR), the optimal length was $L = 189$ ($L_\mu = 0.6047$, Figure 3.14c and 3.14d). Comparing $L = 100$ and $L = 140$ (Figure 3.15) highlights the method's benefits: $L = 100$ produces clear periodic fault impulses with prominent BPFI and harmonics (Fig 3.15b and 3.15d), while $L = 140$ shows random impulses and less-defined BPFI/harmonics (Fig 3.15e and 3.15f). This demonstrates the importance of determining the optimal filter length.

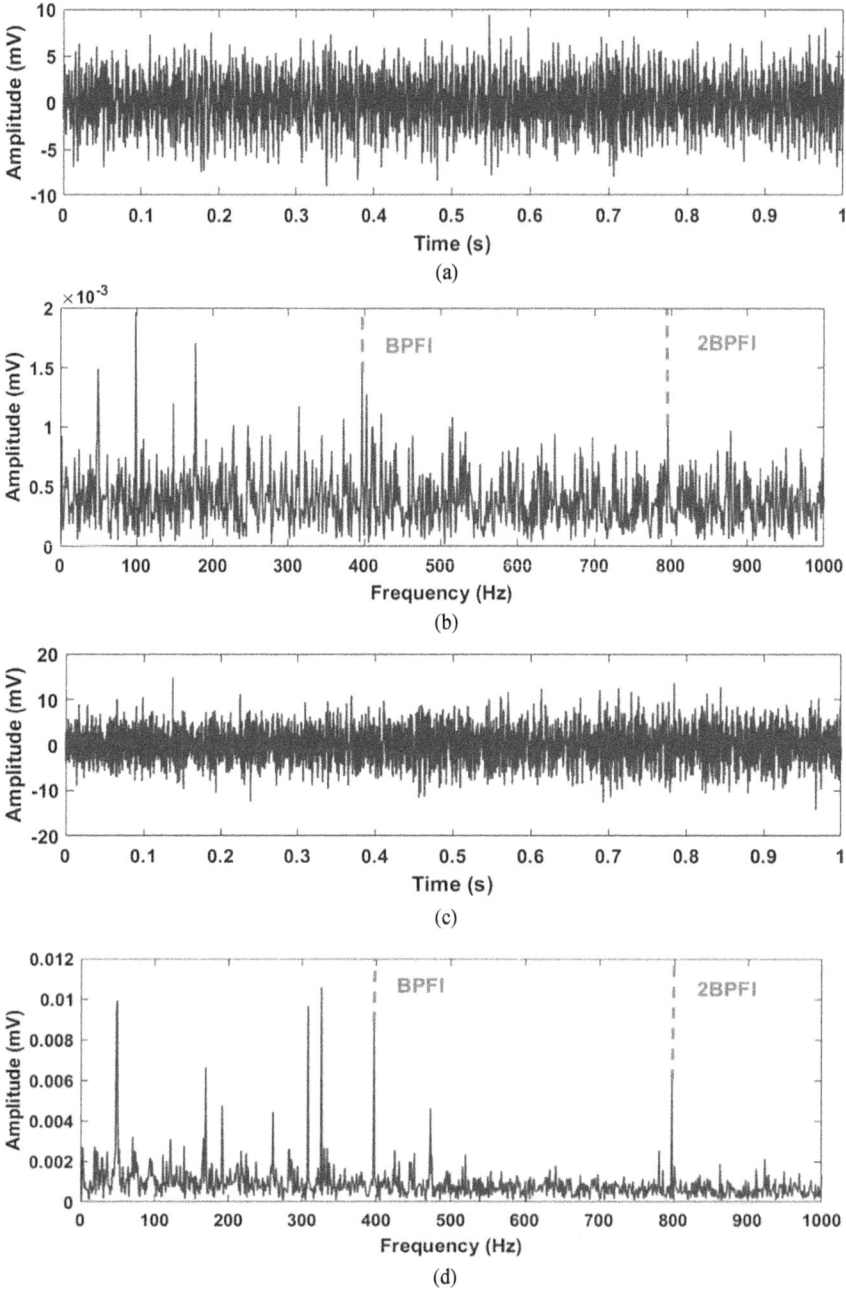

FIGURE 3.13 Time-domain waveform and envelope spectra: (a) waveform of one spall defect at inner race (1-IR), (b) envelope spectrum of 1-IR, (c) waveform of two spall defects at inner race (2-IR), and (d) envelope spectrum of 2-IR.

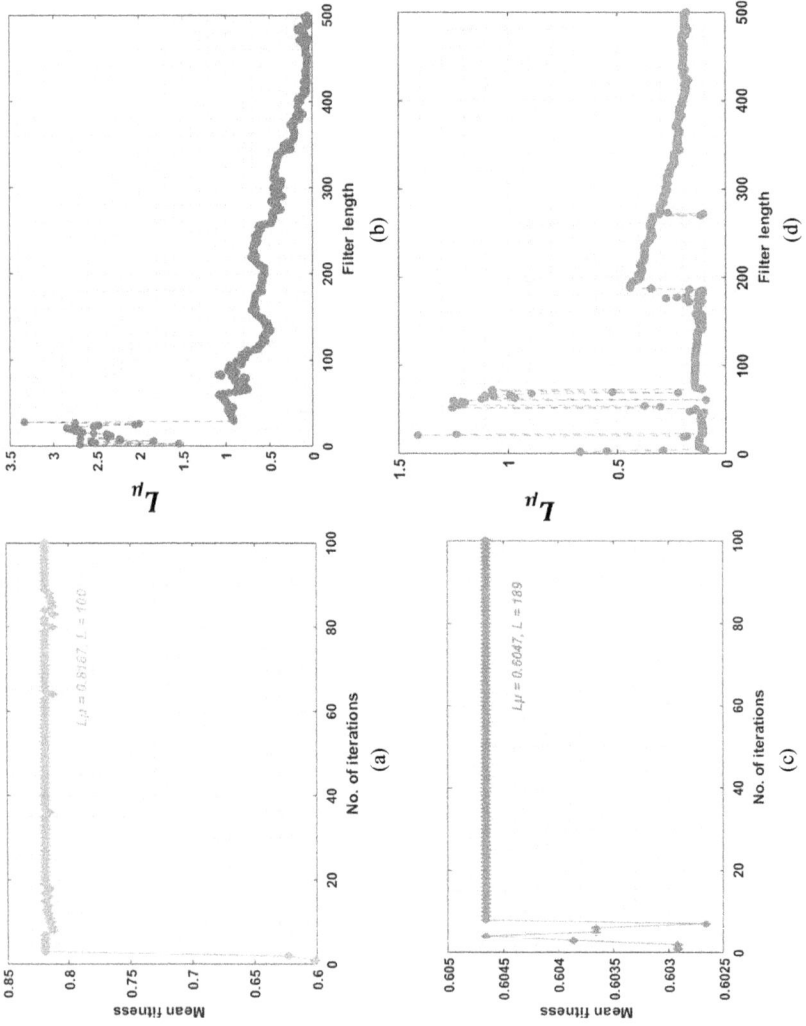

FIGURE 3.14 Selection of optimal filter length for inner race defect. (a) Results of Aquila optimizer for 1-IR optimization, (b) value of L_μ for 1-IR, (c) results of Aquila optimizer for 2-IR optimization, and (d) value of L_μ for 2-IR.

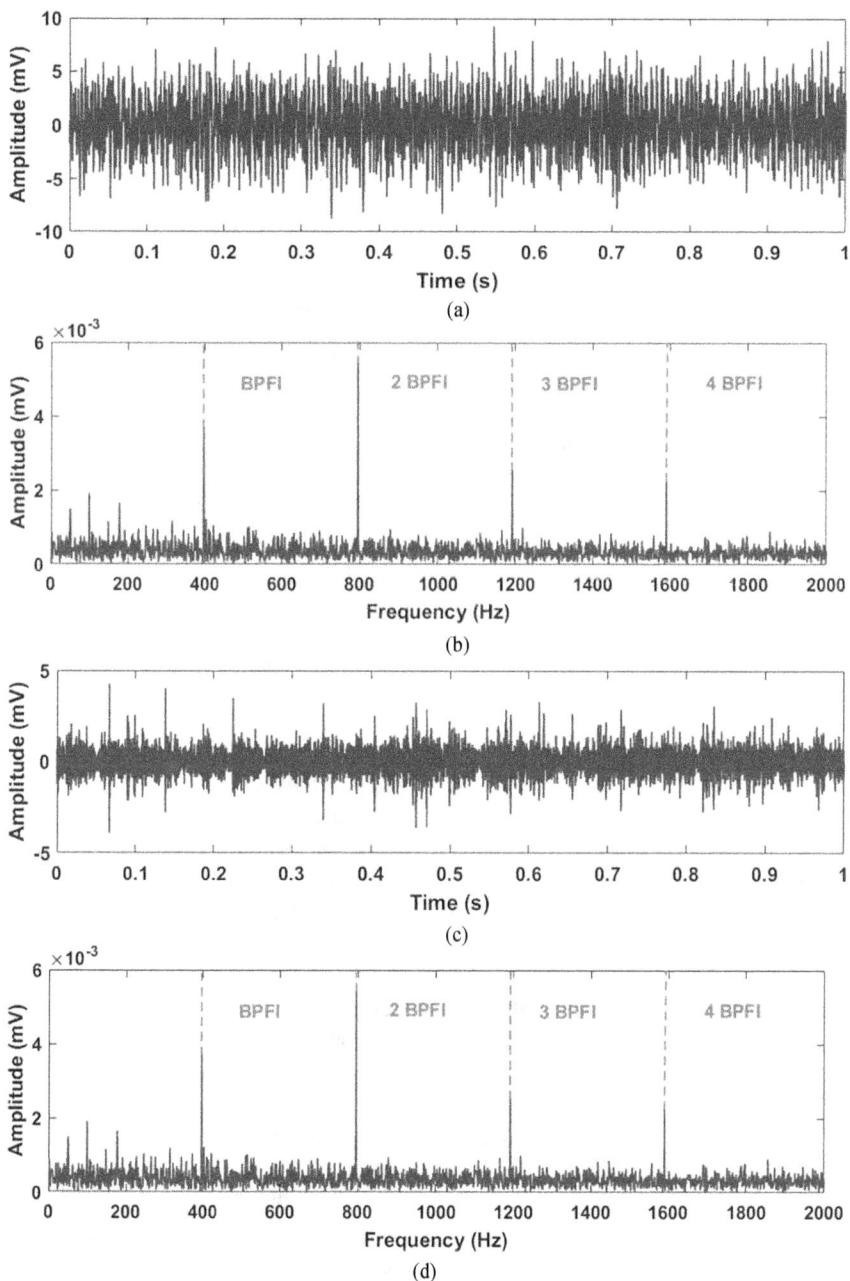

FIGURE 3.15 Results of inner race faults. (a) Time-domain signal, $L = 100$ (optimized) for 1-IR, (b) spectrum of (a), (c) time-domain signal, $L = 189$ (optimized) for 2-IR, (d) spectrum of (c)

(*Continued*)

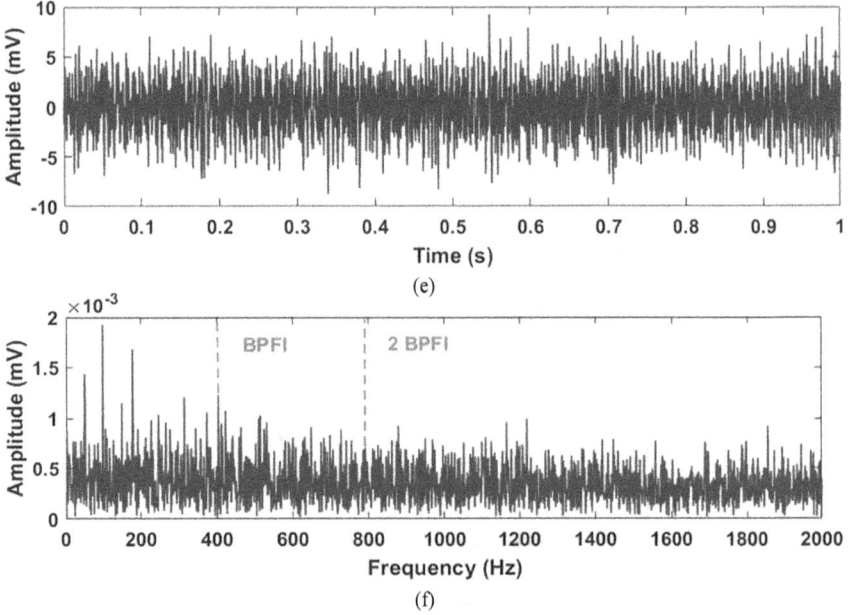

FIGURE 3.15 (**Continued**) Results of inner race faults. (e) time-domain signal, $L = 140$ for 1-IR, and (f) spectrum of (e).

Outer race defects (1-OR and 2-OR) were also analyzed (Figure 3.16). Figures 3.16a and 3.16c shows that periodic fault impulses are masked by significant noise, with diminishing amplitudes. While ball pass frequency of the outer race (BPFO) is visible in Figure 3.16b (1-OR), its harmonics are obscured; for 2-OR, BPFO is not discernible in Figure 3.16d. This suggests that signals from the Francis turbine in harsh operating conditions are significantly weak.

Figure 3.17 shows the results obtained using the AO at the optimized filter length. With an ideal filter length of L = 138, the fitness function L_μ is determined to be 0.2772 for the case of 1-OR, and 0.1421 for the case of 2-OR, with an optimal filter length of L = 159. Figure 3.18 shows the outputs for both outer race faults (1-OR and 2-OR) at the appropriate optimum filter lengths. A continuous recording of impulses is shown in Figure 3.18 (a), (b), (c), and (d). The BPFO, as well as 2BPFO, 3BPFO, and 4BPFO, are clearly evident in the envelope spectrum, which shows rotation modulation. However, as the filter length is increased to L = 110, the findings diverge dramatically. The minimum entropy deconvolution (MED) fails to recover periodic impulses, hence the time-domain signal appears as random impulses. Furthermore, the BPFO and its harmonics are undetectable, as seen in Figure 3.18 (e) and 3.18 (f).

Results from the Francis turbine bearing analysis under various conditions highlight the importance of optimal MED filter length selection for accurate fault detection. The correct length enhances the visibility of periodic impulses;

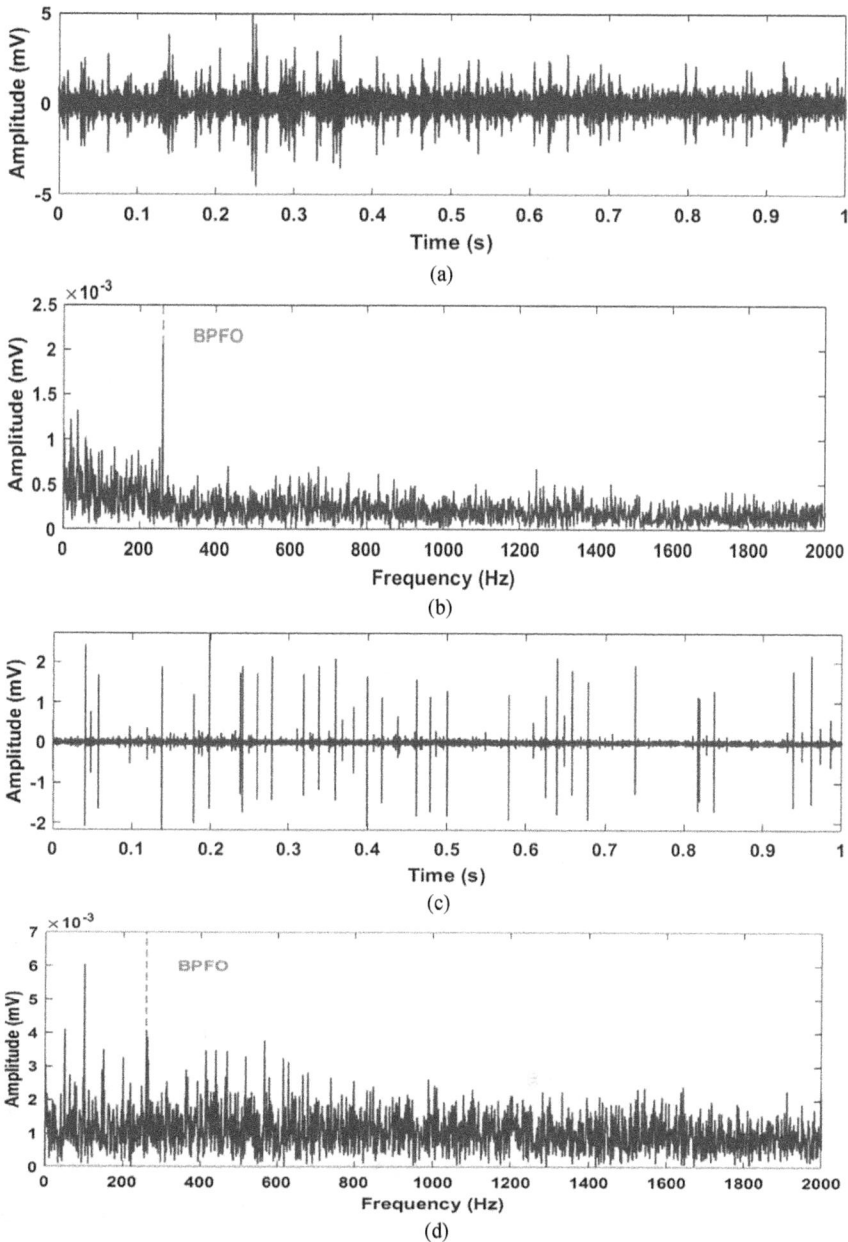

FIGURE 3.16 Time-domain waveform and envelope spectra: (a) one spall defect at outer race (1-OR), (b) envelope spectrum of 1-OR, (c) two spall defects at outer race (2-OR), and (d) envelope spectrum of 2-OR.

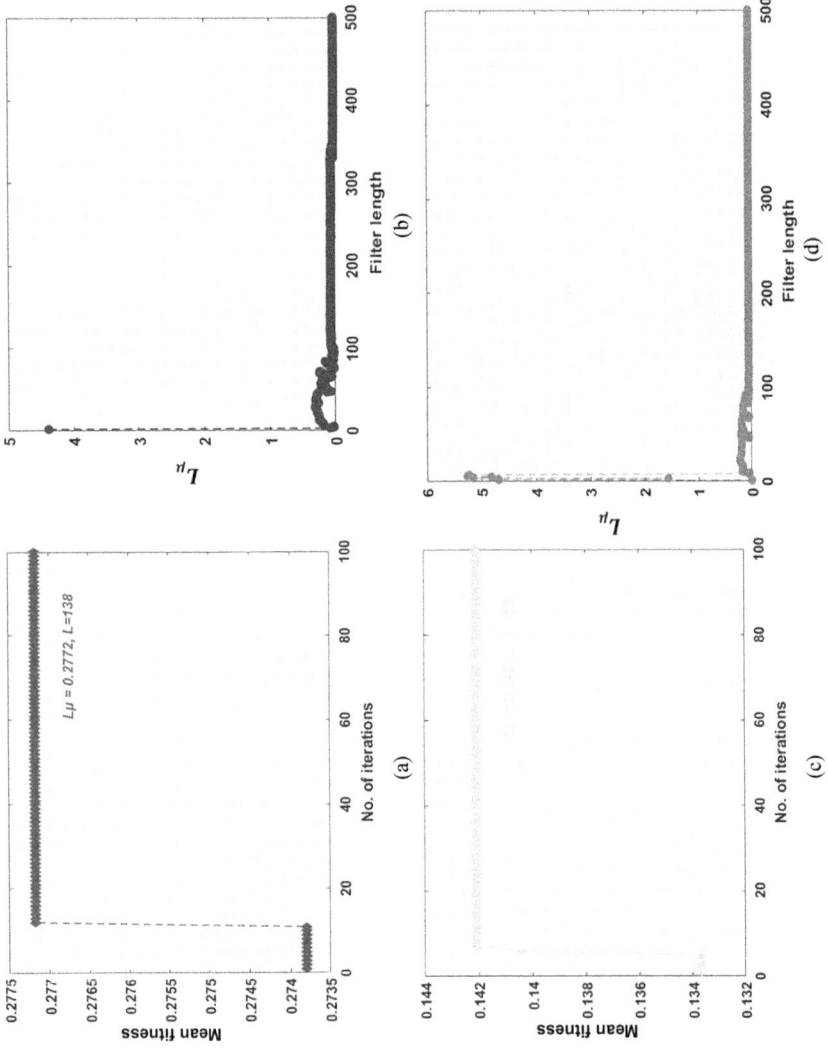

FIGURE 3.17 Selection of optimal filter length for inner race defect: (a) results of Aquila optimizer for 1-OR optimization, (b) value of L_μ for 1-OR, (c) results of Aquila optimizer for 2-OR optimization, and (d) value of L_μ for 2-OR

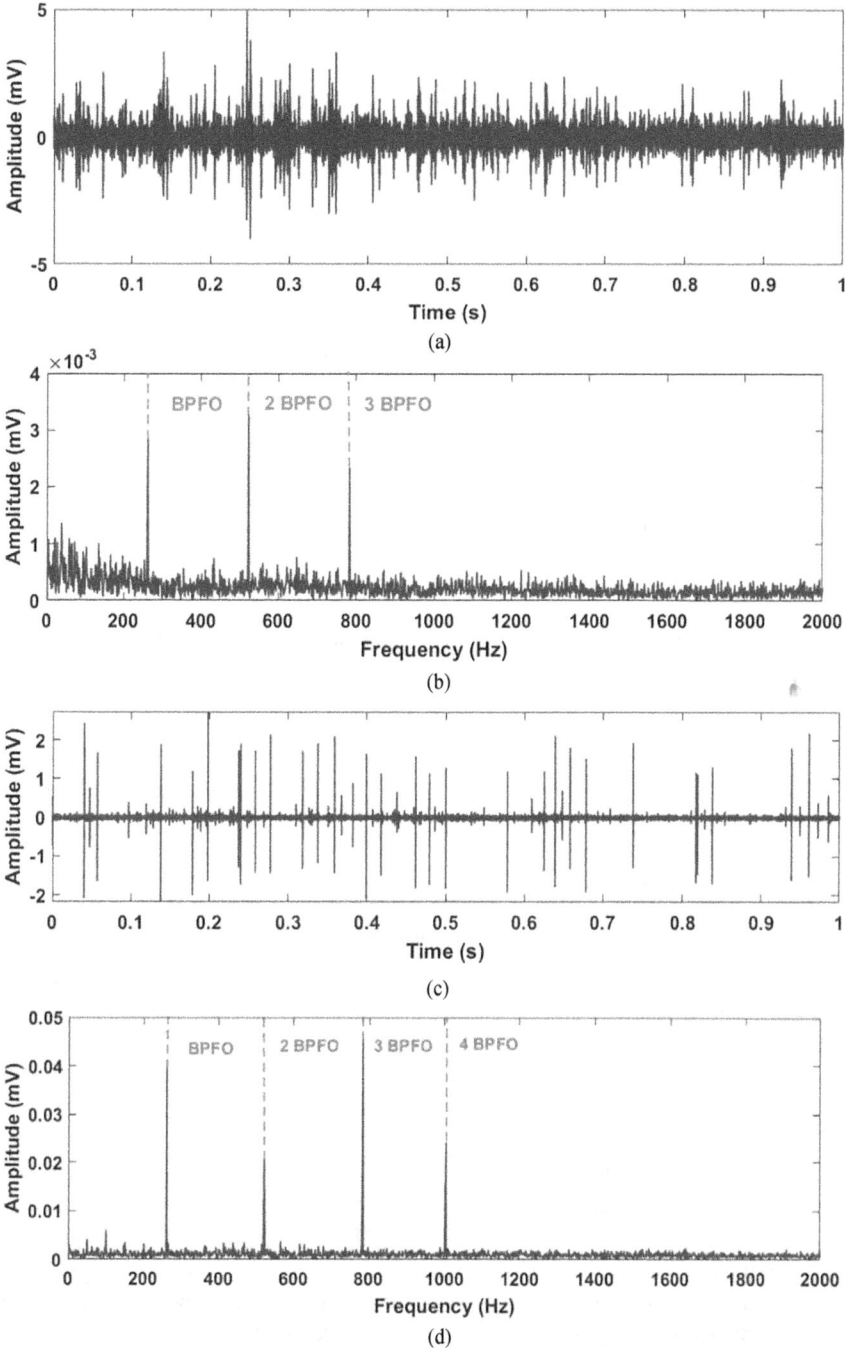

FIGURE 3.18 Results of outer race faults: (a) time-domain signal, $L = 138$ (optimized) for 1-OR, (b) spectrum of (a), (c) time-domain signal, $L = 159$ (optimized) for 2-OR, (d) spectrum of (c)

(*Continued*)

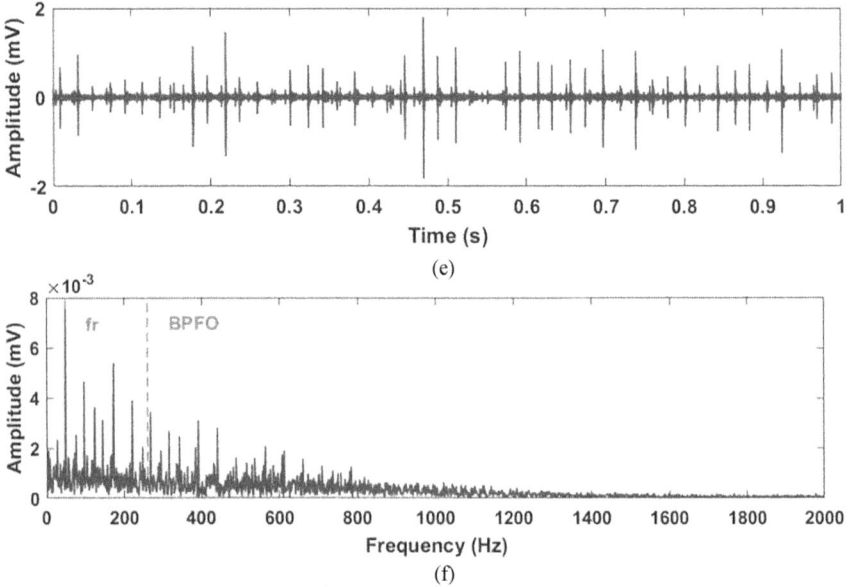

FIGURE 3.18 (**Continued**) Results of outer race faults: (e) time-domain signal, $L = 110$ for 1-OR, and (f) spectrum of (e).

conversely, an incorrect length impairs MED's ability to extract meaningful information from weak signals, hindering accurate fault diagnosis. The proposed method effectively addresses these challenges.

3.4 COMPARISON OF THE PROPOSED METHOD TO EXISTING STATE-OF-THE-ART METHODS

MED performance can be enhanced by: (i) optimizing filter coefficients [85, 86] and (ii) substituting different fitness functions for kurtosis [79–84]. The advantages of the suggested technique are highlighted when compared to PSO-MED and MCKD.

3.4.1 Comparison with PSO-MED

The enhanced MED technique that employs PSO converts the filter coefficients into generalized spherical coordinates, enabling PSO to determine the optimal solution. Reference [85] suggests that PSO-MED is superior to MED, especially in situations with a low signal-to-noise ratio. In the comparative analysis, the parameters for PSO-MED were configured to correspond to those in Reference [85]. Furthermore, in PSO-MED, it is essential to predefine the filter length to accurately detect faults in the bearings of the Francis turbine.

The envelope spectrum for signals associated with inner race defects in bearings, particularly 1-IR and 2-IR, is shown in Figure 3.19 after being processed

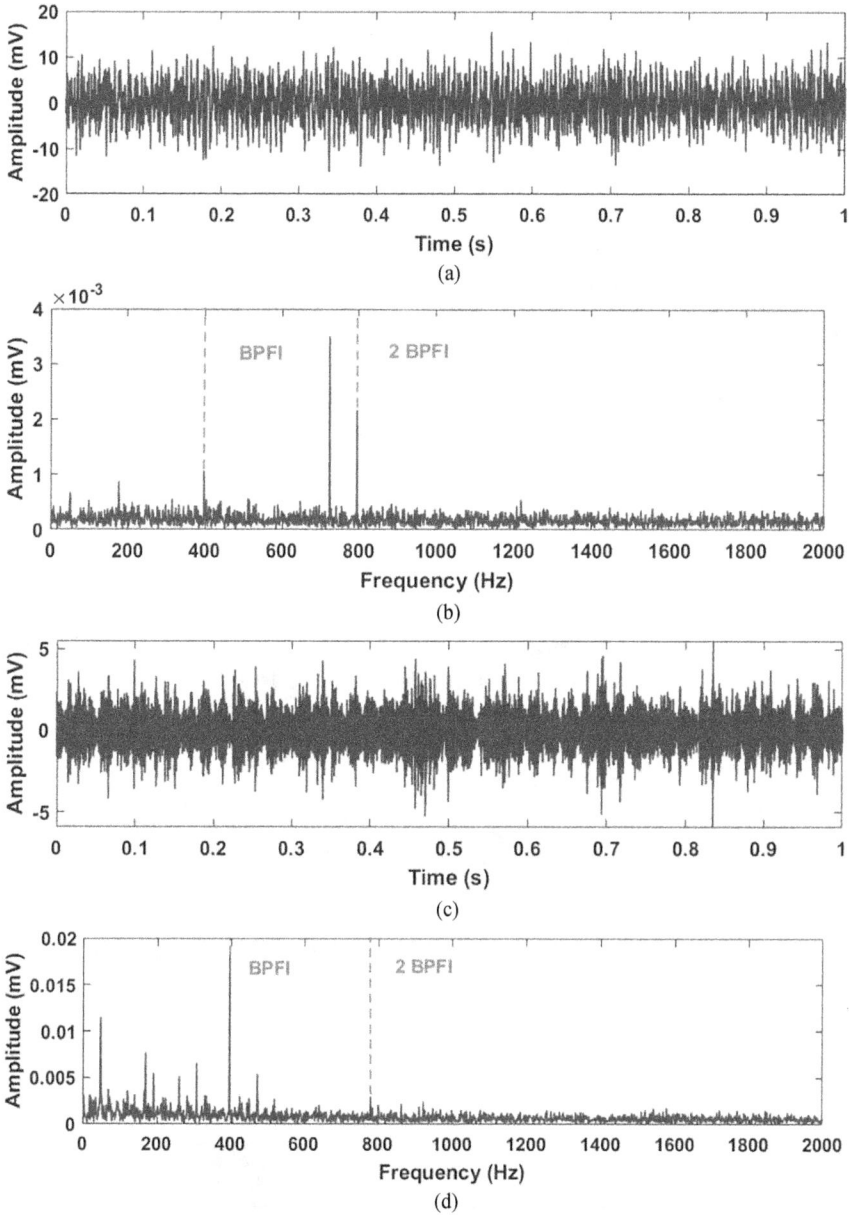

FIGURE 3.19 The results of inner race faults after filtering by PSO-MED: (a) time domain under 1-IR fault, (b) envelope spectrum of (a), (c) time-domain signal under 2-IR fault, and (d) envelope spectrum of (c)

with PSO-MED. This figure illustrates that while the BPFIs and their harmonics are detectable, they are significantly masked by other noise frequencies, leading to less pronounced fault characteristics than those depicted in Figure 3.15. The outcomes of the PSO-MED method applied to outer race defects, specifically both 1-OR and 2-OR, are presented in Figure 3.20, which showcases time waveforms

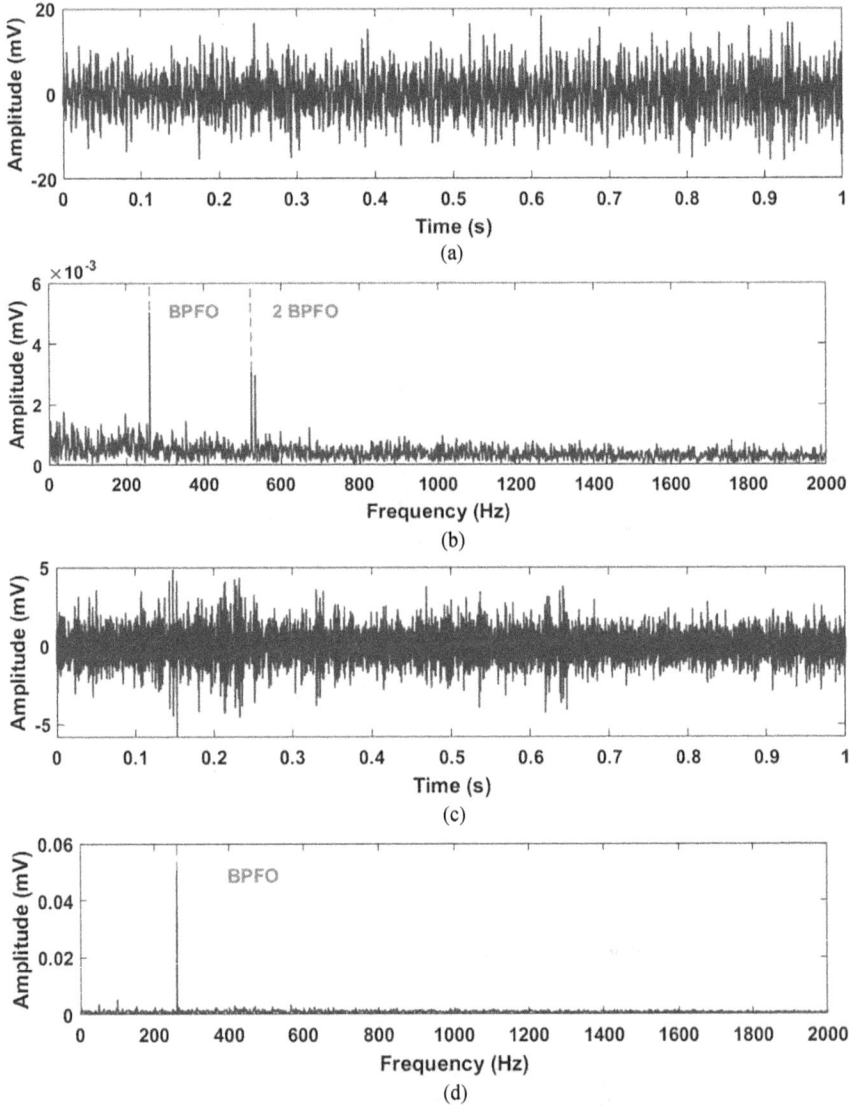

FIGURE 3.20 The results of outer race fault after filtering by PSO-MED. (a) Time-domain signal under 1-OR, (b) envelope spectrum of (a), (c) time-domain signal under 2-OR, and (d) envelope spectrum of (c).

and envelope spectra. Notable impulses can be recognized in the envelope spectrum of Figure 3.20; however, the harmonics are not discernible. These results suggest that PSO-MED does not perform adequately in detecting bearing defects in signals affected by the demanding operating conditions of the Francis turbine.

PSO-MED attempts to optimize filter coefficients by maximizing signal kurtosis using PSO-generated particle sequences to find an "optimal solution." However, the iterative filter coefficient update (Eq. 3.6) is computationally expensive and inaccurate, resulting in poor performance with the turbine's challenging bearing signals.

3.4.2 COMPARISON WITH MCKD

McDonald et al. [97] proposed MCKD, an improved MED method using correlation instead of kurtosis as the fitness function, enhancing fault impulse visibility and outperforming PSO-MED. However, MCKD struggles with harmonic identification. Accurate MCKD results depend critically on knowing the fault period, T [97], which also influences filter length (Eq. 3.13 [85, 88]) and diagnostic accuracy. A better estimate of T generally leads to a longer filter length.

$$L > \frac{2f_s}{f_c} \tag{3.13}$$

Here, f_c denotes the resonant frequency of the fault excitation, while f_s represents the sampling frequency.

As a result, it is essential to choose the right fault period T and to carry out a fair comparison. The value of T is derived from Reference [22] and is calculated using $T = f_s$ / BPFI (or BPFO). Specifically, T is set at 111 for inner race faults and at 88.8 for outer race faults. All signals are filtered utilizing the same optimal filter length and shift order. The envelope spectrum of MCKD with inner race defects for both 1-IR and 2-IR is shown in Figure 3.21. It is clear that MCKD can only identify a single BPFI with a faint peak, without any harmonic presence or rotational modulation. Figure 3.22 illustrates the time-domain signal and envelope spectrum for both 1-OR and 2-OR defects as processed by MCKD. Although the time-domain signal lacks distinct impulses, the envelope spectrum displays BPFO intertwined with noise. A comparison between PSO-MED and MCKD indicates that MCKD performs better for inner race defects, whereas PSO-MED excels with outer race defects. Nevertheless, the proposed method outperforms MCKD in detecting periodic impulses with notable amplitude and in extracting BPFI and BPFO harmonics.

Table 3.3 compares the proposed method, PSO-MED, and MCKD in terms of percentage error and computation time. The proposed method shows lower error and computation time than both PSO-MED and MCKD. While conventional MED, PSO-MED, and MCKD only detected defect frequencies, the proposed method identified both defect frequencies and their harmonics.

The proposed algorithm was executed using MATLAB R2019a software, while LabVIEW 2020 was employed as the interface for data acquisition. The

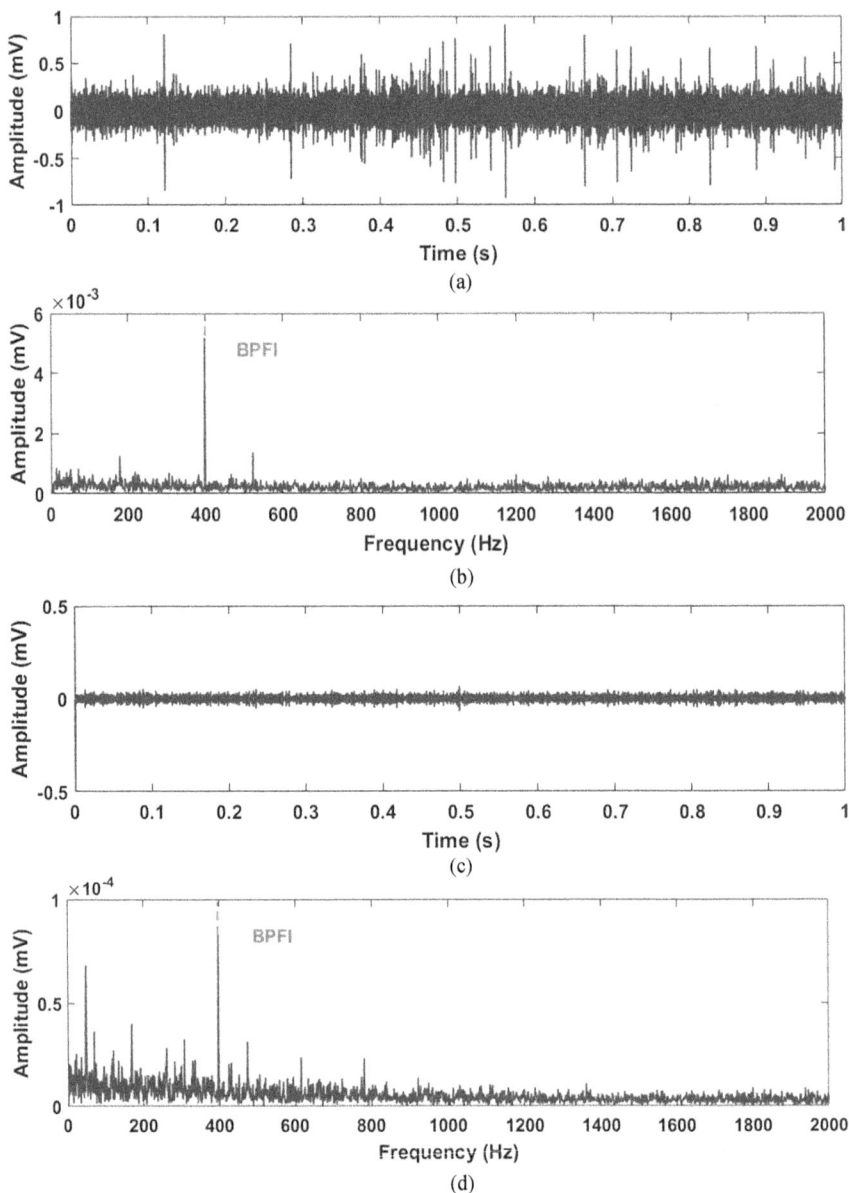

FIGURE 3.21 The results of the inner race fault after filtering by MCKD. (a) Time-domain signal under 1-IR fault, (b) spectrum of (a), (c) time-domain signal under 2-IR fault, and (d) spectrum of (c).

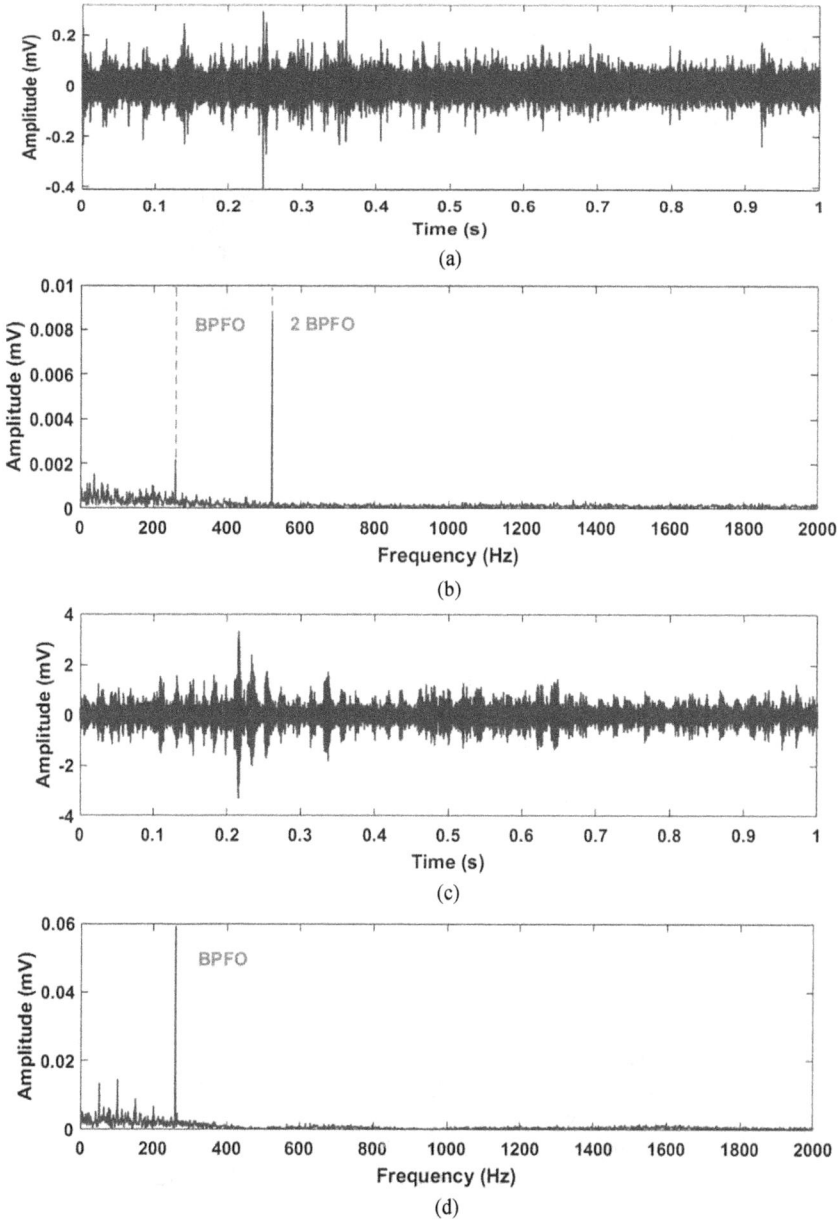

FIGURE 3.22 The results of outer race fault after filtering by MCKD. (a) Time-domain signal under 1-OR, (b) spectrum of (a), (c) time-domain signal under 2-OR, and (d) spectrum of (c).

TABLE 3.3

Comparison of Results of the Proposed Fault Identification Scheme with Other Existing Schemes

Signal Processing Scheme	Defect Type	RPM	Fault Frequency Obtained after Processing Signal (Hz)	Percentage Error in Fault Frequency (%)	Computational Time (sec)
Conventional MED	One seeded defect on the inner race (1-IR)	3040	400	0.6220	20.12
	Two seeded defects on inner race (2-IR)	3040	399	0.3705	17.27
	One seeded defect on outer race (1-OR)	3040	255	2.3729	21.21
	Two seeded defects on outer race (2-OR)	3040	256.12	1.9441	19.84
PSO-MED	One seeded defect on inner race (1-IR)	3040	397	0.1325	18.54
	Two seeded defects on inner race (2-IR)	3040	396.12	0.3539	17.63
	One seeded defect on outer race (1-OR)	3040	260.52	0.2595	17.82
	Two seeded defects on outer race (2-OR)	3040	261.14	0.0222	15.42
MCKD	One seeded defect on inner race (1-IR)	3040	396.5	0.2583	19.73
	Two seeded defects on inner race (2-IR)	3040	397.92	0.0988	17.97
	One seeded defect on outer race (1-OR)	3040	260.01	0.4548	17.13
	Two seeded defects on outer race (2-OR)	3040	262	0.3070	16.57
Improved MED using auto correlation energy and Aquila optimizer	One seeded defect on inner race (1-IR)	3040	397.5	**0.0067**	**14.41**
	Two seeded defects on inner race (2-IR)	3040	397.52	**0.0017**	**13.92**
	One seeded defect on outer race (1-OR)	3040	261.197	**0.0003**	**13.71**
	Two seeded defects on outer race (2-OR)	3040	261.20	**0.0007**	**12.79**

machine was configured with an AMD Ryzen 5 4600 H processor featuring Radeon graphics, operating at 3 GHz and equipped with 8 GB of RAM. It ran on a 64-bit Windows 10 operating system.

The MED is a FIR filter, widely used for blind deconvolution. Setting the filter length correctly is essential before utilizing the MED. Results from simulated and test signals indicate that even a difference of 1 in filter length can yield entirely different outputs, highlighting the significant impact of the chosen filter length on the MED results. Numerous researchers have sought to enhance the performance of the MED filter [79–84], achieving notable results; however, the filter length still needs to be predetermined. An index, (L_μ), has been developed to assess the effect of periodic pulses on the output signal from the MED filter and to minimize the influence of single pulses as much as possible. Findings from various bearing health conditions of the Francis turbine demonstrated that using an optimized filter length can effectively mitigate the adverse effects of a noisy environment.

The superiority of the suggested method can also be demonstrated using an index referred to as the characteristic frequency of the envelope (C_{f_e}), which is effective in distinguishing between various fault levels. Previous literature has utilized similar indexes to evaluate the capabilities of various methods, as referenced in [83–86]. The characteristic frequency of the envelope (C_{f_e}) can be calculated using the following equation:

$$C_{f_e} = \frac{\frac{1}{M} \cdot \sum_{i=1}^{M} A(i \cdot f_{fault})}{\frac{1}{N} \cdot \sum_{i=1}^{N} A(f_i)} = \frac{N \cdot \sum_{i=1}^{M} A(i \cdot f_{fault})}{M \cdot \sum_{i=1}^{N} A(f_i)} \qquad (3.14)$$

Here, $A(f)$ at frequency f in the envelope spectrum represents the amplitude. N stands for the spectrum's spectral lines, M for the multiple of f_{fault}, which in this study is set to 3, and f_{fault} for the fault frequency. With the search bandwidth covering the range $[-5, 5]$, the value of N is set at 500. As shown in Figure 3.23, the characteristic frequency of the envelope C_{f_e} of the suggested method has been contrasted with that of the PSO-MED and MCKD methods. This comparison led to the following conclusions:

a. The severe vibration environment that exists during Francis turbine operation greatly reduces the bearing's fault characteristics under a variety of health circumstances. The 1-IR, or spall defect at one point on the inner race, is the least prominent of the several fault modes, whereas the 2-OR, or spall faults at two sites on the outer race, is the most noticeable.

b. The suggested approach successfully addresses the difficulties presented by the highly vibrating environment and extracts the fault characteristics of faulty bearings, which are frequently hidden by a lot of noise. According to the comparison, the suggested approach performs better than PSO-MED and MCKD in detecting the signals connected to the Francis turbine's faulty bearings.

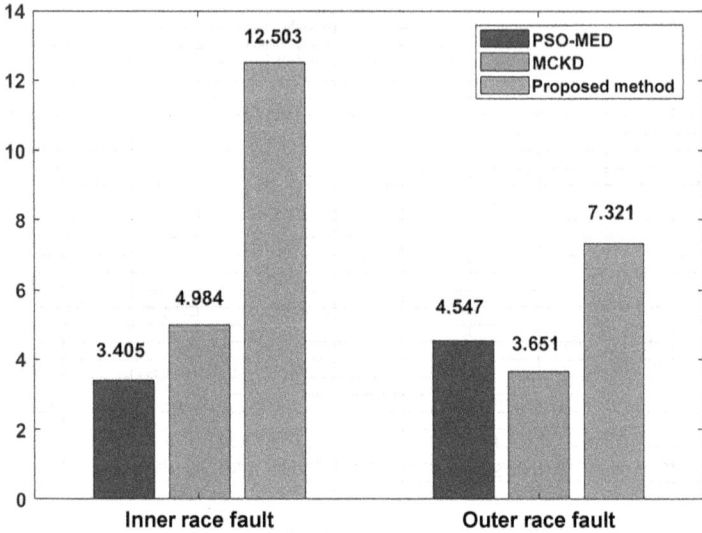

FIGURE 3.23   Output of C_{f_e} index for failure modes.

3.5   CONCLUSION

This research introduces an innovative fault detection method utilizing an AO to enhance the MED filter's performance. The approach employs autocorrelation energy as a fitness function, enabling optimal filter length determination to accurately identify periodic impulses. The technique has been successfully applied to Francis turbine-bearing acoustic signals, where challenging conditions and noise often obscure fault indicators. The proposed method effectively amplifies weak periodic impulses and has been benchmarked against PSO-MED and MCKD. It accurately detects fault frequencies for various bearing defects (1-IR, 2-IR, 1-OR, and 2-OR) with minimal percentage errors (0.0067, 0.0017, 0.0003, and 0.007, respectively) and efficient computation times (14.41, 13.92, 13.71, and 12.79 seconds, respectively). Comparative analysis demonstrates the superiority of this technique in extracting subtle defect characteristics over existing methods.

3.6   SUMMARY

This chapter focuses on the study that used sound signals to investigate bearing issues in the Francis turbine. To accomplish this, a measurement index called autocorrelation energy has been created. This index serves as a fitness function when the MED filter length is optimized using the AO. The suggested technique helps to strengthen the weak periodic impulses that are present in the turbine's difficult operating circumstances.

4 Fault Diagnosis of the Centrifugal Pump

4.1 INTRODUCTION

Hydraulic pumps play a crucial role in various sectors, including agriculture, industry, and domestic applications [36, 98, 99]. Centrifugal pumps, in particular, require careful monitoring to prevent mechanical failures during continuous operation. These pumps can experience three main types of faults: mechanical-induced, system-related, and operational. Mechanical faults involve component failures such as impeller damage [100], bearing issues [101, 102], and shaft misalignment [103]. System faults are associated with improper installation and leakage, while operational faults include blockages, cavitation, and erosion [104, 105]. This research concentrates on mechanical-induced faults, specifically those affecting impellers and bearings. Condition-based monitoring (CBM) is employed to evaluate the health of rotating components by analyzing measurable parameters. When these parameters deviate from normal behavior or decline, the diagnostic system is activated, prompting maintenance actions once predetermined thresholds are exceeded. For rotating machinery, CBM typically involves monitoring vibration, acoustic emission, sound, or pressure signals. In this study, the focus is on vibration-based condition monitoring to diagnose impeller and bearing defects in centrifugal pumps.

4.2 DIAGNOSIS OF IMPELLER DEFECT IN CENTRIFUGAL PUMP (CASE 1)

A centrifugal pump generates a pressure head through its rotating impeller and involute casing. The impeller is attached to a shaft that is supported by bearings. As the impeller rotates, it creates velocity, which is then transformed into pressure by the casing. Various factors can lead to malfunctions in the impeller, including corrosion caused by reactive chemicals, erosion due to solid slurry particles, metallurgical defects, cavitation, and lack of proper lubrication [99, 106].

The vibration signals collected for monitoring may contain noise, which needs to be eliminated for an accurate fault analysis. To find centrifugal pump problems, researchers have been actively examining sophisticated signal processing and machine learning approaches [101–103]. For example, Azizi et al. [107] classified the findings using a generalized regression neural network (GRNN) after utilizing empirical mode decomposition (EMD) to determine the degree of cavitation in centrifugal pumps. The genetic algorithm support vector machine (GA-SVM) model was created by Kumar and Kumar [108] to categorize different flaws using features

DOI: 10.1201/9781003614821-4

that were taken from the scale marginal integration (SMI) signal as well as the raw signal. Although the genetic algorithm showed promise in fault prediction, it had drawbacks, including a sluggish rate of convergence and the ability to become stuck in local minima. Additionally, Kumar et al. [109] employed symmetric cross-entropy of neutrosophic sets to diagnose defective bearings in an axial pump.

Variational mode decomposition (VMD), introduced by Dragomiretskiy, breaks down a signal into intrinsic modes, with the center frequency being calculated in real time, allowing the extracted modes to synchronize accordingly [110]. He demonstrated that VMD outperforms EMD in tone detection and separation. Using a cross-entropy measurement index, Kumar et al. [68] applied the VMD technique to find centrifugal pump problems. VMD was used by Zhang et al. [105] to break down signals in order to identify bearing problems in a multistage centrifugal pump. Mode number and quadratic penalty factor are two examples of VMD parameters that are commonly determined by experience, which can have a substantial effect on VMD's performance and can lead to erroneous decomposition results. For VMD-based decomposition to be effective, the ideal parameter combination must be determined.

To optimize VMD settings, a variety of optimization strategies have been used. Swarm intelligence (SI) techniques stand out among these because they are modeled after the collective behavior of various creatures, including schools of fish, ants, and flocks of birds, and are inspired by natural events [111]. Gravitational search algorithms (GSA), particle swarm optimization (PSO), sine cosine algorithms (SCA), hybrid genetic algorithms and particle swarm optimization (HGPSO), ant colony optimization (ACO), whale optimization algorithms (WOA), grey wolf optimizers (GWO), and grasshopper optimization algorithms (GOA) are a few examples of SI approaches. These methods are used for different optimization problems [50, 112].

This chapter presents ASSA that utilizes opposition and position updating to develop an adaptive VMD technique, applied to vibration signals for detecting defects in centrifugal pumps. By efficiently expanding the starting population size, the proposed technique not only speeds up convergence but also cuts down calculation time. This approach reduces the likelihood that the algorithm would stall when attempting to find the optimal set of VMD decomposition parameters. The sensitive mode for faulty feature extraction is chosen using the weighted kurtosis index. Additionally, the extracted features are ranked using the Pearson correlation coefficient (PCC) technique, which reduces data redundancy and determines each feature's impact on the signal. An extreme learning machine (ELM) is then trained using the chosen characteristics to assess testing and training accuracy.

4.2.1 THEORETICAL BACKGROUND

4.2.1.1 Variational Mode Decomposition (VMD)

VMD is a signal processing method that breaks down the raw vibration signal into intrinsic modes u_k, which maintains a sparsity with the original signal. Each mode is expected to be centered around a specific frequency w_k. The corresponding sparse-mode u_k is selected as the bandwidth in the frequency domain, as described in [110, 113]. To determine the mode bandwidth, the following steps are executed: (1) each mode, u_k, is derived using the Hilbert transform to generate its frequency spectrum, (2) the frequency spectrum of each mode is exponentially shifted to the "baseband" in relation to its central frequency, and (3) the bandwidth of the frequency is then estimated using the (L^2) norm of the gradient. This decomposition process is conducted according to [110].

$$\min \left\{ \sum_k \left\| \partial t \left[\left(\partial(t) + \frac{j}{\pi t} \right) * u_k(t) \right] e^{(iw_k t)} \right\|_2^2 \right\} s.t. \sum_k u_k = f(t) \qquad (4.1)$$

where $f(t)$ denotes an input signal, while $\{u_k\}$ indicates a distinct set of modes and $\{w_k\}$ represents the central frequency. The Dirac distribution ∂t signifies convolution. The penalty factor α serves as the data fidelity constraint, and the Lagrangian multiplier λ is employed to impose this constraint, allowing the optimization problem outlined in Eq. (4.1) to be transformed into an unconstrained form. The augmented Lagrangian $\mathcal{L}$ is expressed in the following equation:

$$\mathcal{L}\left(\{u_k\}, \{w_k\}, \lambda \right) = \alpha \sum_k \left\| \partial t \left[\left(\partial(t) + \frac{j}{\pi t} \right) * u_k(t) \right] e^{-jw_k t} \right\|_2^2 - \left\| f(t) - \sum_k u_i(t) \right\|_2^2$$
$$+ \langle \lambda(t), f(t) - \sum_k u_i(t) \rangle \qquad (4.2)$$

The alternate direction method of multipliers (ADMM) addresses the aforementioned minimization issue (augmented Lagrangian $\mathcal{L}$) to find the saddle point and generates a series of suboptimizations. By leveraging the solutions from these suboptimizations, ADMM directly optimizes the problem in the Fourier domain, as referenced in [110]. The complete algorithm is outlined in [110]. The values of u_k and w_k are updated in accordance with the ADMM optimization problem during processing. Equation (4.2) updates the variational mode function (VMF) concerning u_k, and following this update of the VMF, the suboptimal problem is expressed as follows:

$$u_k^{n+1} = argmin_{u_k \in X} \left\{ \begin{array}{c} \alpha \left\| \partial t \left[\left(\partial(t) + \frac{j}{\pi t} \right) * u_k(t) \right] e^{-jw_k t} \right\|_2^2 \\ + \left\| f(t) - \sum_i u_i(t) + \frac{\lambda(t)}{2} \right\|_2^2 \end{array} \right\} \qquad (4.3)$$

The optimal solution to the quadratic equation presented in Eq. (4.3) can be conveniently determined using the Fourier transform in the frequency domain, with the filter adjusted to the central frequency. Subsequently, Eq. (4.4) updates the values of the modes $\hat{u}_k$.

$$\hat{u}_k^{n+1}(w) = \frac{\hat{f}(w) - \sum_{i \neq k} \hat{u}_i(w) + \frac{\hat{\lambda}(w)}{2}}{1 + 2\alpha(w - w_k)^2} \tag{4.4}$$

Filtering is performed on the current residual in conjunction with the signal prior to applying $1/(w - w_k)^2$ using a Wiener filter, and the mode u_k is updated in accordance with Eq. (4.4). Hermitian symmetric completion is used to derive the spectrum of each VMF (u_k). The real part of the inverse Fourier transform of the filtered signal yields the modes u_k in the time domain. Eq. (4.2) is optimized with respect to w_k to ensure that the central frequency does not affect reconstruction fidelity. The relevant problem is referenced from [110].

$$w_k^{n+1} = argmin_{w_k} \left\{ \alpha \left\| \partial t \left[\left(\partial(t) + \frac{j}{\pi t} \right) * u_k(t) \right] e^{(jw_k t)} \right\|_2^2 \right\} \tag{4.5}$$

Upon resolving the aforementioned suboptimization problem, the result for the central frequency is obtained as follows:

$$w_k^{n+1} = \frac{\int_0^\infty w |\hat{u}_k(w)|^2 dw}{\int_0^\infty |\hat{u}_k|^2 dw} \tag{4.6}$$

The central frequency w_k is updated based on the center of gravity of the power spectrum of the corresponding mode, as indicated in Eq. (4.6).

The aforementioned equations indicate that four parameters—namely, mode number (K), quadratic penalty factor (α), tolerance (τ), and convergence criterion (ϵ)—are essential for the VMD procedure and must be defined ahead of time. The parameters τ and ϵ typically use their default values from the original VMD, as they have minimal impact on the decomposition outcomes. However, the mode number K should not be predetermined without prior knowledge of the signal being analyzed, as its appropriateness cannot be assessed, which may affect the accuracy and efficiency of the decomposition results. The quadratic penalty factor α helps suppress noise interference in the signal and regulates the frequency bandwidth, necessitating careful selection. Therefore, finding the optimal combination of these parameters is crucial for VMD and serves as the motivation for this work.

4.2.1.2   Extreme Learning Machine (ELM)

Huang et al. [114] introduced the ELM for applications in both regression and classification. The ELM is built on a single-layer feedforward network (SLFN), as illustrated in Figure 4.1. It comprises three layers: the input layer, the hidden layer, and the output layer. For M arbitrary samples $(X_j, t_j) \in R^n X R^m$, the SLFN with L hidden nodes and an activation function $G(\alpha_i, \beta_i, X_i)$ is mathematically represented as shown in Eq. (6.7) [114]:

$$f_L(X_j) = \sum_{i=1}^{L} \beta_i G(\alpha_i . X_j + b_i) = t_j, j = 1, \ldots, M. \tag{4.7}$$

Here, α_i and b_i are the learning parameters associated with the hidden nodes, where α_i connects the weight vector of the input nodes to the ith hidden node, and b_i represents the threshold for the ith hidden node. Additionally, β_i signifies the output weight, and t_j indicates the test point, while the activation function $G(\alpha_i, \beta_i, X_i)$ produces the output for the ith hidden node. This equation can also be expressed in matrix form as follows:

$$H\beta = T \tag{4.8}$$

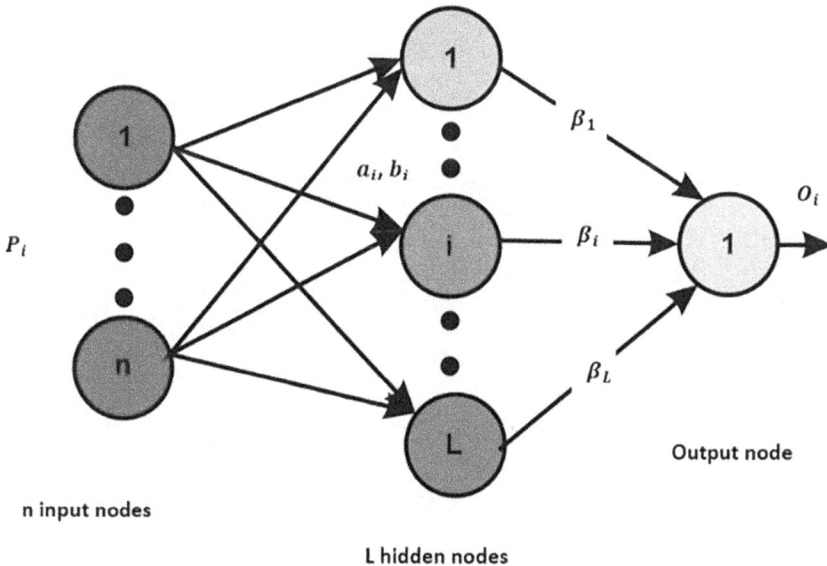

FIGURE 4.1   Extreme learning machine (ELM) structure.

where

$$H = \left(\alpha_1,\ldots,\alpha_L,b_1,\ldots,b_L,X_1,\ldots,X_N\right) = \begin{bmatrix} G(\alpha_1,\beta_1,X_1) & \cdots & G(\alpha_L,\beta_L,X_1) \\ \cdot & & \cdot \\ \cdot & \cdots & \cdot \\ \cdot & & \cdot \\ G(\alpha_1,\beta_1,X_N) & \cdots & G(\alpha_L,\beta_L,X_N) \end{bmatrix}_{N \times L} \tag{4.9}$$

and

$$\beta = \begin{bmatrix} \beta_1^T \\ \cdot \\ \cdot \\ \cdot \\ \beta_L^T \end{bmatrix}, T = \begin{bmatrix} T_1^T \\ \cdot \\ \cdot \\ \cdot \\ T_N^T \end{bmatrix} \tag{4.10}$$

According to ELM theories, the values of α_i and b_i are assigned randomly for all hidden nodes rather than being adjusted through tuning. The solution to the preceding equation is estimated by

$$\beta = H^* T \tag{4.11}$$

where H^* represents the inverse of the output matrix H, and the Moore–Penrose generalized inverse is employed for this purpose. The procedures included in the ELM algorithm are summarized as follows:

Step 1: The hidden nodes' learning parameters, α_i and b_i, are assigned at random.

Step 2: It is necessary to compute the hidden layer's output matrix H.

Step 3: The Moore–Penrose generalized inverse is used to calculate the inverse of the hidden layer output matrix H.

4.2.2 PROPOSED SCHEME

To enhance the analysis of vibration signals for pump fault detection, a VMD process is proposed. This adaptation is achieved by optimizing the VMD's parameters, specifically the mode number K and the quadratic penalty factor, through a specialized search rule. This rule relies on two primary criteria: a measurement index to evaluate the quality of the VMD results and an efficient search method. A refined salp swarm algorithm (SSA), incorporating opposition-based learning and position updating, is developed as the search method to effectively identify the optimal parameter combination. Furthermore, to accurately locate the sensitive mode within the decomposed signal, a new measurement index, weighted kurtosis, is introduced. This index improves upon traditional kurtosis by incorporating the correlation coefficient and considering density distribution, thus providing a more comprehensive evaluation. Following the identification of the sensitive mode, relevant features are extracted and utilized by an ELM to enable automated fault detection in the pump.

4.2.2.1   Weighted Kurtosis Index

The measurement index is a crucial component in making VMD adaptive, as it assesses the effectiveness of the decomposition results. Previous research has indicated that kurtosis and correlation coefficients are two significant indices for diagnosing faults in rotating machinery using vibration signals [115]. The kurtosis index primarily depends on the density distribution of the impacts caused by faults [116]. Utilizing maximum kurtosis alone to optimize VMD parameters can be problematic because it may fail to detect impactful events with high amplitudes if their density distribution is spread out. While the correlation coefficient offers a rapid assessment of signal similarity, its vulnerability to noise, particularly in signals from faulty components, limits its reliability [115]. To overcome the limitations of relying solely on maximum kurtosis or the correlation coefficient, a combined metric, the weighted kurtosis index, is proposed. This hybrid index will function as the fitness function, guiding the optimization process to determine the most effective VMD parameters [117]. The following Eq. (4.12) represents the weighted kurtosis index (KCI).

$$KCI = KI.|C| \tag{4.12}$$

where KI represents the kurtosis index for input signal $x(n)$ and is expressed as

$$KI = \frac{\frac{1}{N}\sum_{n=0}^{N-1} x^4(n)}{\left(\frac{1}{N}\sum_{n=0}^{N-1} x^2(n)\right)^2} \tag{4.13}$$

where the length of the signal is given by N. Considering $E[.]$ a mathematical expectation, the correlation C between x and y is expressed as

$$C = \frac{E\left[(x-\bar{x})(y-\bar{y})\right]}{E\left[(x-\bar{x})^2\right]E\left[(y-\bar{y})^2\right]} \tag{4.14}$$

4.2.2.2   Ameliorated Salp Swarm Algorithm for Optimizing VMD Parameters

Mirjalili et al. [118] introduced the SSA, an optimization method rooted in SI. This technique emulates the foraging patterns of salps or chains of salps in the deep ocean. Within SSA, the initial random population is divided into two groups:

leaders and followers, facilitating the creation of the mathematical model of the salp swarm [118]. The leader directs the salp chain, while the followers move behind the leaders. The basic principles of SSA and the Enhanced Salp Swarm Algorithm (ASSA) proposed in this study are detailed in the following subsections, along with explanations for the modifications made to SSA. Shared processes such as population initialization, function evaluation, and swarm division are relevant to both SSA and ASSA. To improve SSA's convergence speed, opposition-based learning is utilized. The approach for updating the positions of both leaders and followers is refined through various equations, which are detailed in the relevant subsections to thoroughly develop ASSA.

A. Initialization of population
   To begin, the population is randomly generated within the search space using a uniform distribution, as demonstrated in Eq. (4.15).

$$x_{ij} = x_j^{\min} + r_{ij}\left(x_j^{\max} - x_j^{\min}\right); \left(i = 1, 2, \ldots, NP; j = 1, 2, \ldots, D\right) \qquad (4.15)$$

   where NP represents the number of populations, D signifies the dimension of the search space, $x^{\min}$ denotes the lower bound of the search space, $x^{\max}$ indicates the upper bound, and r_{ij} is a uniformly generated random number within the range (0,1).

B. Opposition-based learning
   Nature-inspired optimization algorithms typically begin with random initial guesses for potential solutions across a defined search space. However, this random initialization can lead to lengthy computation times. To mitigate this, an approach known as opposition-based learning can be employed. Instead of solely relying on random guesses, each initial solution is paired with its "opposite" counterpart, and the fitness of both is evaluated. The superior solution, whether the original random guess or its opposite, is then selected as the starting point. This strategy, by initializing with closer approximations and validating them against the fitness function, effectively reduces computation time and accelerates convergence. This technique is applied consistently across the initial population and has been integrated into the fundamental SSA, where the population initialization utilizes this opposition-based learning framework.

$$x_{O_{ij}} = x_j^{\max} + x_j^{\min} - x_{ij} \qquad (4.16)$$

   where $x_{O_{ij}}$ represents the salp population from opposition-based learning.

C. Function evaluation
   The fitness of the swarm is assessed using Eq. (4.16). Subsequently, the optimal function value, $F_b(i)$, is derived mathematically and expressed as

$$F_b(i) = \min\left\{F\left(x(i)\right)\right\} \tag{4.17}$$

where $i = 1, 2, \ldots, NP$

The best salp position is saved corresponding to the best function value, $F_b(i)$.

D. Dividing the swarm

The entire swarm is segmented into two groups known as leaders and followers. The proportion of leaders can range from 10% to 90%, as indicated in [118]. In this research, the leaders and followers are divided equally, meaning they are assigned the same percentage.

E. Update the position of the leader

Similar to other swarm-based optimization methods, the position of each salp serves as a candidate solution stored in a two-dimensional matrix referred to as X for an m-dimensional search problem, where m represents the number of design variables. Within the search space, the optimal position corresponds to the best food source, denoted as F. The positions of the leaders are updated according to Eq. (4.18).

$$X_j^1 = \begin{cases} F_j + c_1\left(\left(ub_j - lb_j\right)c_2 + lb_j\right), c_3 \geq P \\ F_j - c_1\left(\left(ub_j - lb_j\right)c_2 + lb_j\right), c_3 < P \end{cases} \tag{4.18}$$

where X_j^1 denotes the position of the leader (first salp), P is the probability used to determine the leader's position, and F_j represents a food source in the jth dimension. ub_j and lb_j are the upper and lower bounds, respectively, for the jth dimension. The variables c_1, c_2, and c_3 are random values, with c_1 playing a crucial role in balancing exploitation and exploration. The definition of c_1 is provided in Eq. (4.19) as follows:

$$c_1 = 2e^{-\left(\frac{4k}{L}\right)^2} \tag{4.19}$$

where k represents the current iteration and L denotes the maximum number of iterations. The random variables c_2 and c_3 are generated uniformly within the range of 0 and 1. The probability P is defined in Eq. (4.20) as follows:

$$P = \tanh\left|S(i) - DF\right| \tag{4.20}$$

where $S(i)$ represents the fitness of x_{ij}, while DF signifies the best fitness achieved across all iterations. Eq. (4.19) has been introduced as a modification that facilitates the updating of the leaders' positions.

F. Update the position of the followers
The follower position is updated according to Eq. (4.20), which is based on
Newton's Law of Motion.

$$X_j^i = \frac{1}{2}\left((1-\alpha)X_j^i + \alpha X_j^{i-1}\right)$$ (4.21)

where $2 \le i \le L$, X_j^i indicates the position of ith follower in the salp chain of jth dimension. And α is the weighting factor defined in Eq. (4.22)

$$\alpha = rand\left(-\text{var}1, \text{var}1\right)$$ (4.22)

where

$$\text{var}1 = atanh\left(-\frac{1}{\max L} + 1\right)$$ (4.23)

Eqs. (4.22) and (4.23) are incorporated in Eq. (4.21) as modifications in the basic salp swarm which updates the follower's position.
    Pseudocode of the ASSA Algorithm is presented in the following manner.

Initialize the salp population NP1 x_{ij} using Eq. (4.15).
Apply opposition-based learning on this initial population to get NP2 members.
Calculate objective function on NP1 and NP2 using Eq. (4.25).
Select best NP members out of (NP1+NP2).
Select members, xbest(j) with best function value and designate as gbest.
while $(l \le L)$
Calculate the objective function of each individual.
Update c_1 by Eq. (4.19)
        for *i = 1: size of salp population*
            if *i <= half of the salp population*
            generate c_2 and c_3 randomly within [0,1]
            calculate the value of P using Eq. (4.20)
            update the position of the leaders using the Eq. (4.18)
                else if *i > half of the salp population*
                generate the variable var1 using Eq. (4.23)
                create α using Eq. (4.22)
                update the position of the follower using Eq. (4.21)
                end
        end
    updated salp position
    end

amend the salps based on upper and lower bounds of variables.
evaluate the objective function on the updated salp position
update the value xbest and gbest
end

4.2.2.3 Pearson Correlation Coefficient-Based Feature Ranking

In the context of large datasets, filter-type algorithms are typically employed to rank the features. A feature filter is essentially a function of correlation or information that returns the relevant index $J(S \mid D, C)$ [119]. This index evaluates the relevance of a given feature subset (S) for the task (C), which can involve either classification or approximation of the data D. The relevant index is computed for each individual feature X_i, where $i = 1 \ldots N$, to establish the ranking order $J\left(X_{i_1}\right) \leq J\left(X_{i_2}\right) \ldots \leq J\left(X_{i_N}\right)$. Features with the lowest ranks are typically filtered out. The PCC is one method used to determine the rank of features based on correlation measurement [119]. The PCC is expressed as follows:

$$e(X,C) = \frac{E(XC) - E(X)E(C)}{\sqrt{\sigma^2(X)\sigma^2(C)}} = \frac{\sum_i (x_i - \bar{x}_i)(c_i - \bar{c}_i)}{\sqrt{\sum_i (x_i - \bar{x}_i)^2 \sum_j (c_j - \bar{c}_j)^2}} \quad (4.24)$$

where X represents a feature with value x, and class C is a class containing values c. If $e(X,C)$ is ± 1, then X and C are considered dependent; conversely, if $e(X,C)$ is zero, X and C are uncorrelated. The error function $P(X \sim C) = erf\left(|e(X,C)|\sqrt{N/2}\right)$ is utilized to calculate the probability that two variables are correlated. Decreasing values of the error function $P(X \sim C)$ indicate feature ranking and organize the feature list accordingly.

4.2.3 FAULT IDENTIFICATION APPROACH

This chapter introduces a parameter-adaptive VMD technique, driven by an ASSA. The ASSA utilizes the maximum weighted kurtosis, as defined in Eq. (4.25), as its fitness function to optimize VMD parameters. To improve the algorithm's performance, opposition-based learning is integrated into the original SSA, leading to the development of ASSA. Since the optimization aims to minimize the fitness function, the maximization problem inherent in the weighted kurtosis is transformed into a minimization problem by negating the fitness function, as shown in Eq. (4.25). Furthermore, the study proposes a feature ranking method based on the PCC to determine feature relevance for fault diagnosis. Finally, these ranked features are used to train and test an ELM for accurate fault detection.

$$\begin{cases} objective\ function = \min_{\upsilon(k,\alpha)} \left(-KCI_i\right) \\ s.t, k = 2,3,\ldots,7 \\ \alpha\ \varepsilon\left[1000,\ldots,10,000\right] \end{cases} \quad (4.25)$$

where KCI_i (for $(i = 1, 2, ..., K)$) represents the weighted kurtosis for the decomposition modes of VMD. The parameters $\upsilon(k, \alpha)$ are those of VMD that need optimization. The mode number k varies within the interval [2, 7], while the quadratic penalty factor α takes values within the range [1000, 10000]. The parameter ranges have been determined based on a comprehensive literature review.

The detailed steps of the proposed methodology are outlined below:

- The acquired vibration signal is the input into VMD using the specified ranges of parameters that require optimization. The value of the objective function is stored for each iteration.
- Initialize the parameters of the ASSA using a population size of N and setting the maximum number of iterations to L.
- Retrieve the modes following the decomposition of the vibration signal using VMD. Next, calculate the objective function for each mode.
- If $l \geq L$, then the desired condition is met, and the iteration concludes. Otherwise, increment l by 1 and continue the iteration.
- Maintain the set of ideal parameters determined by the goal function's lowest value. Furthermore, keep track of the sensitive mode, which is the mode linked to the highest weighted kurtosis index.
- The PCC is used to rate the distinguishing traits when they are taken out of the sensitive state. After that, the data is saved.
- To assess the model's accuracy during testing and training, the resulting data is fed into the ELM.

A flow chart of the process is given in Figure 4.2.

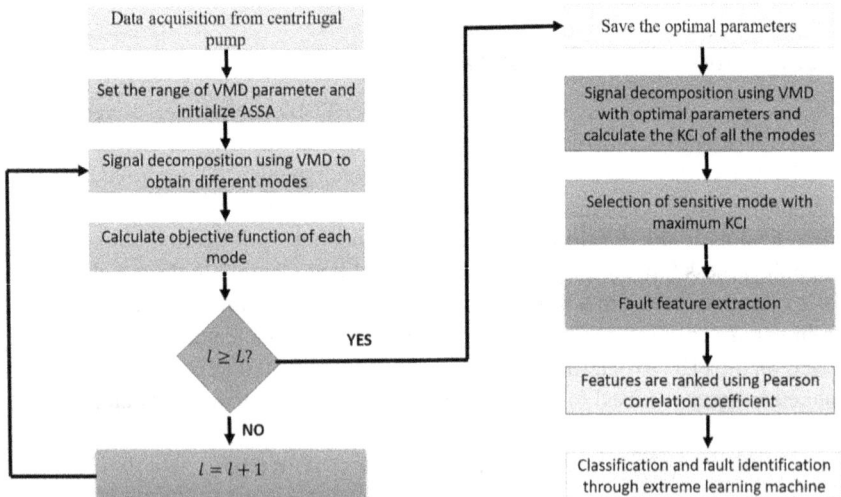

FIGURE 4.2 Flowchart for adaptive VMD method for fault identification in the centrifugal pump.

4.2.4 EXPERIMENTATION

4.2.4.1 Test Rig

The centrifugal pump test rig is the source of the experimental dataset. The schematic diagram and visual representation of the pump test rig are shown in Figure 4.3(a) and 4.3(b), respectively. The pump runs at a speed of 2800 rpm or 46.67 Hz. Table 4.1 contains the pump's detailed specifications. Two bearings support the

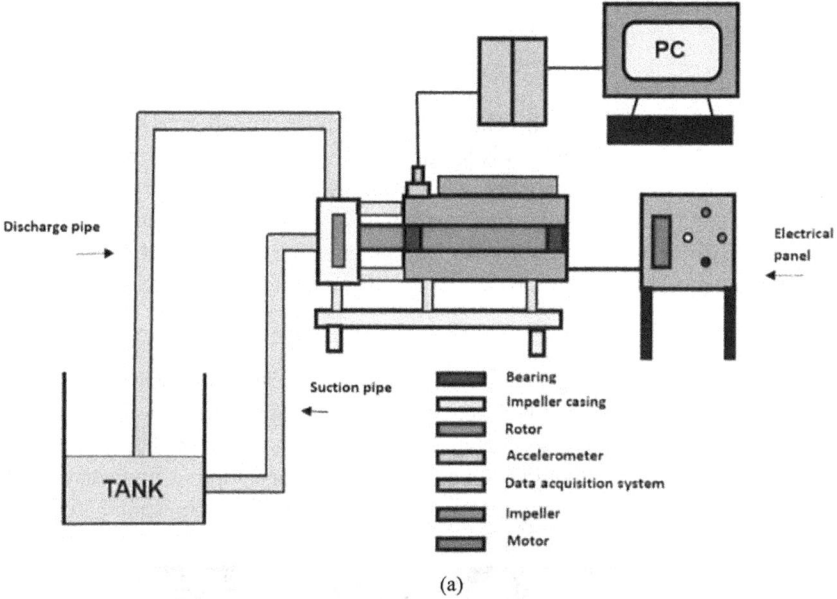

(a)

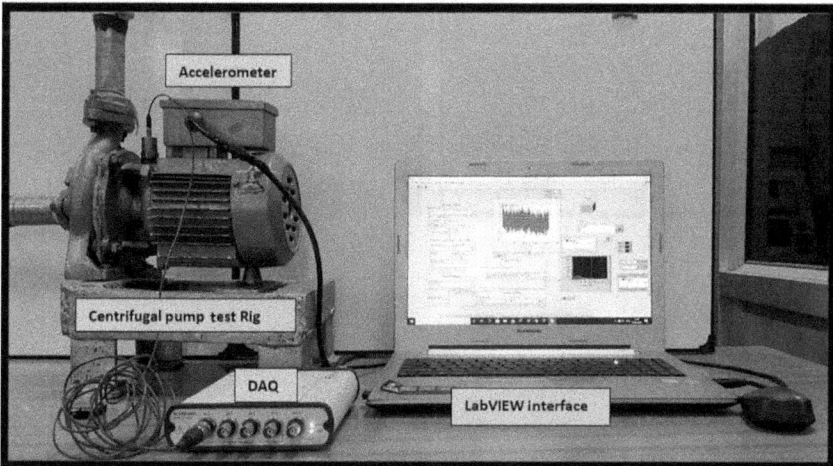

(b)

FIGURE 4.3 (a) Schematic of centrifugal pump test rig and (b) a typical photograph of centrifugal pump test rig with an accelerometer placed for data acquisition.

TABLE 4.1

Specification of Centrifugal Pump

Power supply	230/240 V
Motor power	0.5 Hp
Discharge	1.61 litre/s
Impeller type	Closed
Impeller diameter	118.88 mm
Impeller vanes	3

pump shaft: Bearing 1, which is located closer to the impeller and is designated 6203-ZZ, and Bearing 2, which is located farther away from the impeller and is designated 6202-ZZ. The impeller, which is housed in a casing and positioned on the rotor shaft, has vanes that, when rotating, pull water axially through the impeller's eye. By doing this, the water is given kinetic energy, which allows it to flow outward radially through the casing and transform it into potential energy, or head.

4.2.4.2 Data Acquisition

The vibration signals are collected using a uniaxial accelerometer with a sensitivity of 100 mV/g, which is mounted near the impeller casing, as depicted in Figure 4.3(b). A National Instruments 24-bit, 4-channel data acquisition (DAQ) system is utilized to capture the vibration signals, operating at a sampling frequency of 70 kHz. 7000 data points, covering 0.1 seconds, are evaluated. As shown in Figure 4.4,

(a)   (b)

(c)

FIGURE 4.4   Different operating conditions: (a) clogging, (b) blade cut, and (c) wheel cut.

the study is conducted under a variety of impeller settings. The adaptive VMD approach is used to process the raw signal obtained from the centrifugal test rig. The modified salp swarm technique is used to optimize the two main parameters of VMD, the mode number K, and the quadratic penalty factor α. Other VMD parameters are used as recommended in [110]. The ideal set of VMD parameters is determined using the suggested ASSA algorithm. With these ideal pairings, the maximum weighted kurtosis is used to determine the most important mode, which is then processed for additional examination.

First, data is collected for a pump with a normal (defect-free) impeller installed that runs at 2800 rpm (corresponding to an operational frequency of 46.67 Hz). In every impeller situation under investigation, the pump keeps its speed constant. Figure 4.5(a) displays the raw time-domain signal for the impeller condition without defects. After that, this signal is converted into the frequency domain, as shown in Figure 4.5(b), where the 47 Hz characteristic frequency which corresponds to the operating frequency of the pump is emphasized. Under all health conditions, the pump runs at a constant rpm, and hence the Fast Fourier Transform (FFT) related to the operating frequency is constant. The adaptive VMD approach, which is based on the ASSA, breaks down the raw signal into different modes. The quadratic penalty factor α and the mode number K are found to be 3 and 1000, respectively, based on the ASSA. Figure 4.5(c) provides illustrations of the various modes. Every mode's weighted kurtosis is determined; the third mode yields the highest weighted kurtosis value, 12.73, and is chosen for feature extraction. Twenty signals in all are examined in the following scenarios: wheel cuts, clogging, blade cuts, and no defect.

Similar to the analysis for the defect-free condition, adaptive VMD is also used to deconstruct the data gathered under blocked impeller conditions into several modes. ASSA optimizes this process by applying a quadratic penalty factor α of 4000 and a mode number K of 3. The three options in this case produce weighted kurtosis values of 1.95, 3.08, and 1.99; the third mode, with the greatest weighted kurtosis value, is selected for additional processing. Figure 4.6(a) shows the time-domain signal, Figure 4.6(b) shows the frequency-domain signal, and Figure 4.6(c) shows the decomposed modes.

Figure 4.7(a) displays the blade cut impeller condition's raw signal, and Figure 4.7(b) displays the matching frequency domain signal. For this signal, the optimal K and α parameter values are found to be 3 and 1000, respectively. As illustrated in Figure 4.7(c), the raw signal is broken down into three modes using these parameters. The first, second, and third modes have weighted kurtosis values of 1.62, 1.60, and 8.62, respectively. In this instance, the third mode is also chosen for additional examination because it has the highest weighted kurtosis value.

The wheel-cut impeller situation is handled in the same way. Figure 4.8(a) displays the time-domain signal, and Figure 4.8(b) displays the equivalent frequency-domain signal. It is discovered that 3 and 2000 are the ideal parameter combinations for the mode number and penalty factor, respectively. Figure 4.8(c) shows the three resultant modes. The first, second, and third modes have weighted kurtosis values of 1.54, 3.87, and 11.63, respectively. The prominent mode is determined to be the one with a weighted kurtosis of 11.63.

(a)

(b)

(c)

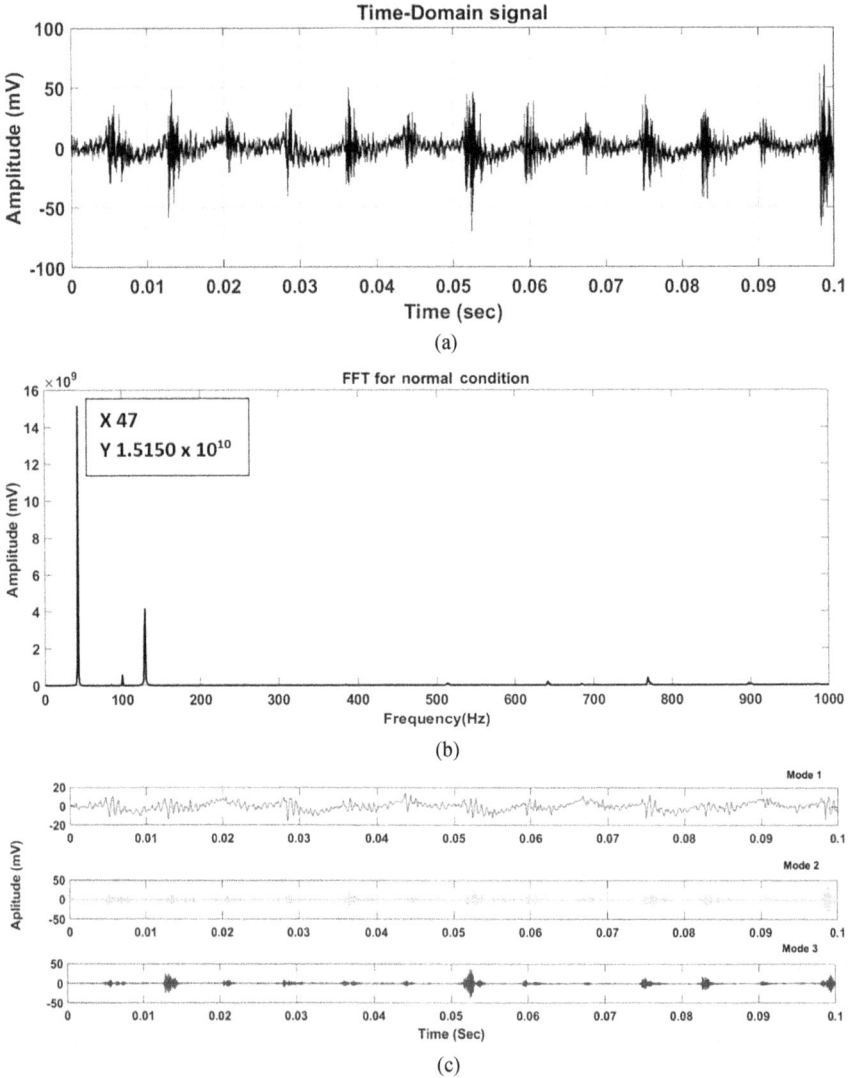

FIGURE 4.5 (a) Raw signal with non-defective impeller. (b) FFT for non-defective impeller. (c) Different modes obtained by applying adaptive VMD to the raw signal.

4.2.4.3 Feature Extraction

A total of 80 prominent modes (20 for each condition: healthy (no defects), clogged, blade cut, and wheel cut impeller conditions) are obtained from the adaptive VMD method based on ASSA, utilizing weighted kurtosis as the measurement index. Subsequently, 11 features are extracted from the prominent modes

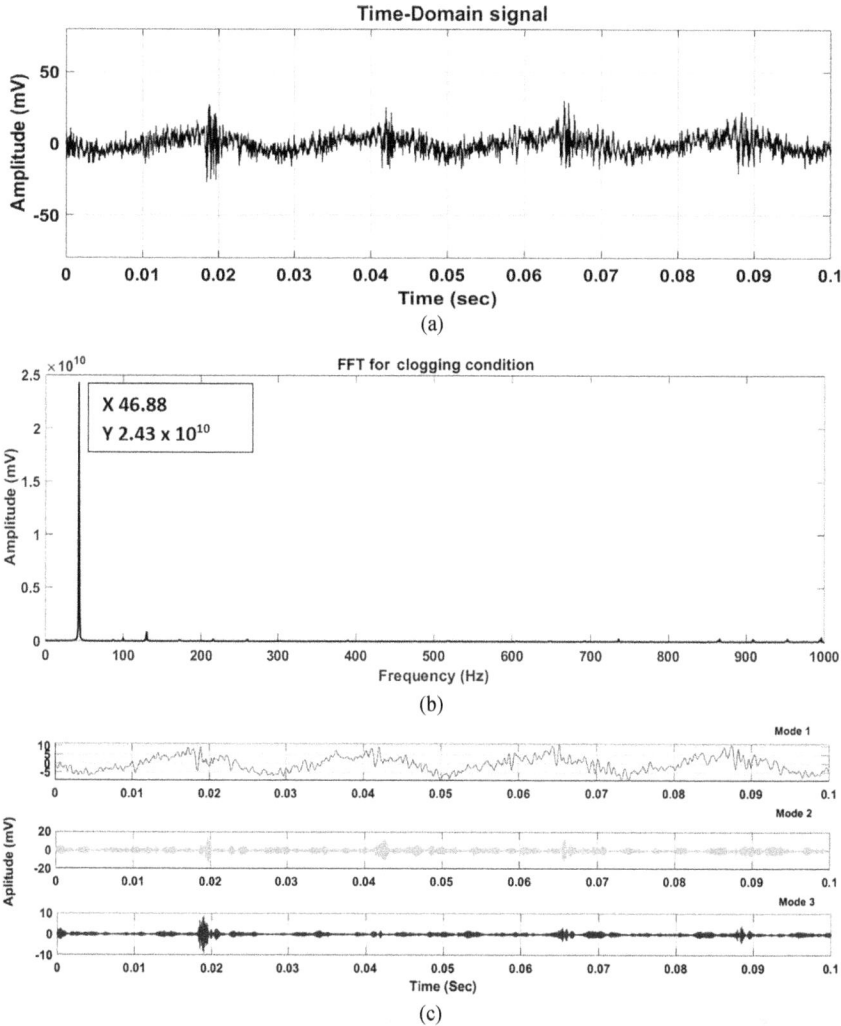

FIGURE 4.6 (a) The raw signal under clogged impeller condition. (b) FFT under clogged impeller condition. (c) Three different modes obtained by applying adaptive VMD under the clogged impeller condition.

decomposed by adaptive VMD. The list of features, along with their definitions, is presented in Table 4.2. The extracted features are normalized within the range of [0, 1] using the following mathematical formula:

$$Normalized\ feature = \frac{Feature - \min(Feature)}{\max(Feature) - \min(Feature)} \quad (4.26)$$

(a)

(b)

(c)

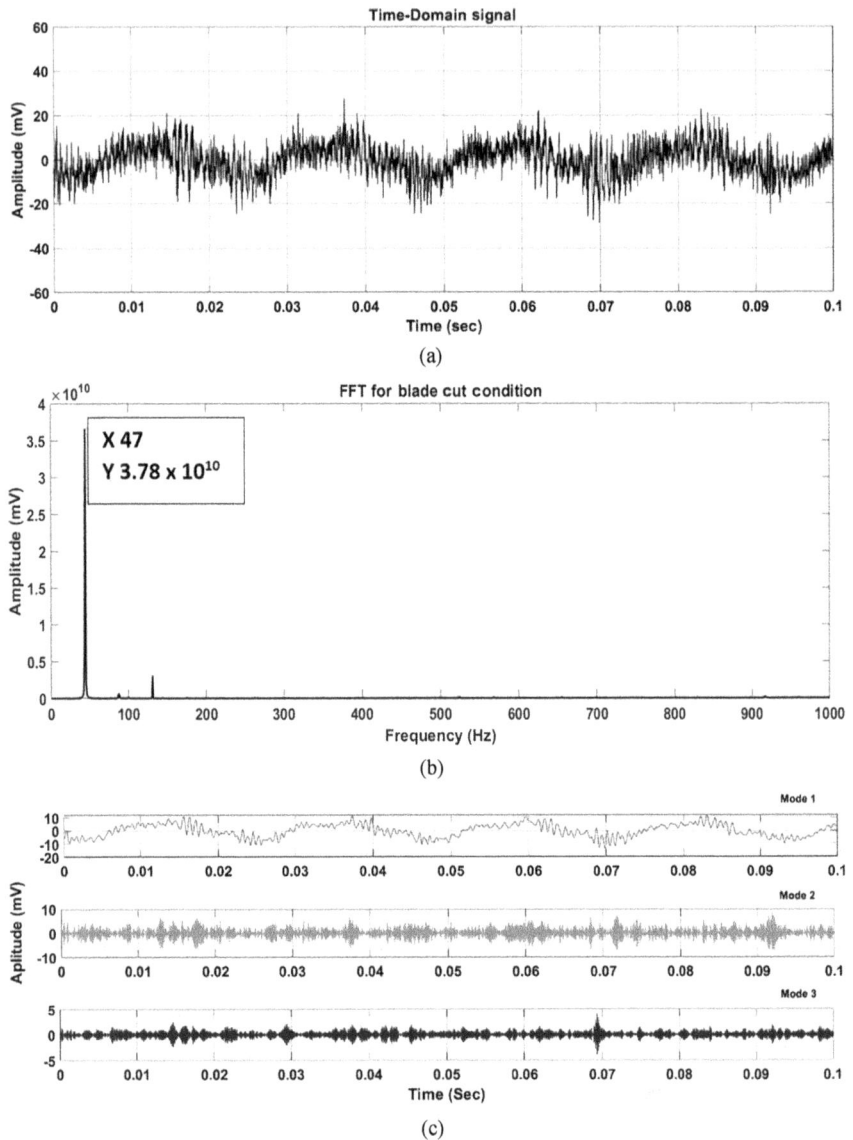

FIGURE 4.7   (a) The raw signal under blade-cut impeller condition. (b) FFT under blade-cut impeller condition. (c) Three modes obtained by applying adaptive VMD under blade-cut impeller condition.

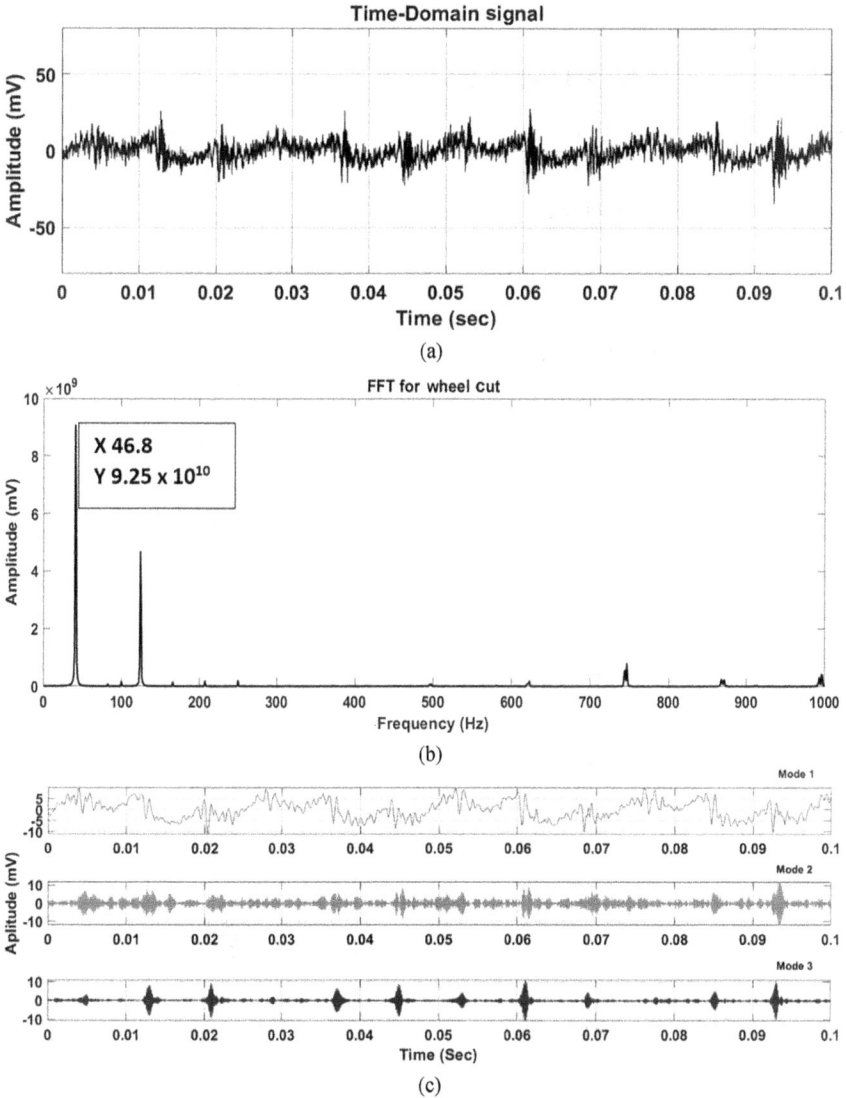

FIGURE 4.8 (a) The raw signal under wheel-cut impeller condition. (b) FFT under wheel-cut impeller condition. (c) Different modes obtained by applying adaptive VMD under wheel-cut impeller condition.

TABLE 4.2
Extracted Features with Their Definition

S. No.	Features	Definitions
1.	Standard Deviation (x_{std})	$x_{std} = \sqrt{\sum_{i=1}^{N} (x(i) - x_m)^2} \Big/ N$
2.	Peak (x_p)	$x_p = \max |x(i)|$
3.	Skewness (x_{ske})	$x_{ske} = \sum_{i=1}^{N} (x(i) - x_m)^3 \Big/ N$
4.	Kurtosis (x_{kur})	$x_{kur} = \sum_{i=1}^{N} (x(i) - x_m)^4 \Big/ Nx_{std}^4$
5.	Root Mean Square (x_{rms})	$x_{rms} = \sqrt{\sum_{i=1}^{N} x(i)^2} \Big/ N$
6.	Peak Factor (PF)	$PF = x_p \Big/ x_{rms}$
7.	Square Root Amplitude (x_{sra})	$x_{sra} = \left(\sum_{i=1}^{N} \sqrt{|x(i)|} \Big/ N \right)^2$
8.	Shape Factor (SF)	$SF = x_{rms} \Big/ \left(\sum_{i=1}^{N} |x(i)| \Big/ N \right)$
9.	Impulse Factor (IF)	$IF = x_p \Big/ \left(\sum_{i=1}^{N} |x(i)| \Big/ N \right)$
10.	Wavelet Packet Decomposition (WPD) Energy	$WPD_i = \sum_{n=1}^{N} |x_i(n)|^2 \Big/ \sum_{k=0}^{2^j-1} \sum_{n=1}^{N} |x_k(n)|^2$
11.	Permutation entropy	$\dfrac{-1}{n-1} \sum_{j=1}^{n!} P_j \log_2 (P_j)$

Here, x represents data; N represents the number of data points; x_m is the average value of x; $x_k(n)$ represents the decomposition coefficient for kth sequence; and j is the level of WPD decomposition.

4.2.5 RESULT AND DISCUSSION

4.2.5.1 Comparison of the ASSA with Other Art of Optimization

The effectiveness of the proposed optimization algorithm (ASSA) is assessed using 23 benchmark functions. ASSA is compared with other advanced optimization algorithms, including SSA, GWO, GOA, SCA, and HGPSO, with metrics such as mean, standard deviation, best, worst, and median values being utilized for comparison. The results are summarized in Table 4.3. ASSA proved

TABLE 4.3
Comparison of the Proposed Algorithm with Other Optimization Algorithms at Benchmark Functions

Function		ASSA (Proposed)	SSA	GWO	GOA	SCA	HGAPSO
F1	Mean	1.1063e–93	9.1209e–09	1.6672e–27	1.5788e–09	2.7425e–25	2.9579e–31
	Standard Deviation	**3.2380e–94**	1.6818e–09	3.2816e–27	1.2994e–09	1.2119e–24	7.3334e–31
	Best	5.5194e–94	6.2194e–09	5.4044e–30	1.8378e–10	5.1600e–35	6.4055e–34
	Worst	1.7057e–93	1.3329e–08	1.4561e–26	5.5621e–10	5.4229e–24	3.3693e–30
	Median	1.0649e–93	9.1999e–09	3.7141e–28	1.1901e–09	1.8613e–29	9.1882e–32
F2	Mean	5.3305e–48	7.7236e–06	7.4550e–17	0.9993	1.1173e–17	7.5501e–248
	Standard Deviation	**1.1164e–48**	2.8604e–06	5.5643e–17	1.5429	3.4603e–17	0.0000
	Best	3.9818e–48	4.7307e–06	1.5632e–17	3.5204e–04	5.1631e–22	5.8965e–301
	Worst	6.8723e–48	1.7809e–05	2.6718e–16	6.4414	1.4762e–16	1.5100e–246
	Median	5.2462e–48	7.1882e–06	6.2314e–17	0.4598	1.2468e–19	4.3475e–295
F3	Mean	1.1245e–93	1.2365e–09	2.4787e–05	1.2172e–07	2.8603e–09	6.1307e–30
	Standard Deviation	**1.1005e–93**	3.8545e–10	7.4077e–05	2.8852e–07	1.1367e–08	1.7521e–29
	Best	1.8863e–94	6.0467e–10	7.3815e–08	1.5850e–10	6.6900e–19	8.6162e–32
	Worst	4.3993e–93	1.9480e–09	3.0164e–04	1.2462e–06	5.1007e–08	7.9903e–29
	Median	7.2784e–94	1.1170e–09	4.9493e–07	1.0015e–08	8.7528e–13	1.1867e–30
F4	Mean	1.1824e–47	1.3264e–05	7.1506e–07	2.6024e–05	3.1534e–08	2.0042
	Standard Deviation	**2.5702e–48**	2.4542e–06	8.1461e–07	1.3879e–05	6.8891e–08	0.9001
	Best	5.6332e–48	8.1067e–06	6.9376e–08	9.0241e–06	1.3787e–07	0.4645
	Worst	1.6194e–47	1.8578e–09	3.3931e–06	5.4687e–05	3.0085e–07	3.5475
	Median	1.2105e–47	1.2999e–05	3.6015e–07	2.2629e–05	6.1902e–09	2.0167

(Continued)

TABLE 4.3 (Continued)

Comparison of the Proposed Algorithm with Other Optimization Algorithms at Benchmark Functions

Function		ASSA (Proposed)	SSA	GWO	GOA	SCA	HGAPSO
F5	Mean	6.9279	12.3668	26.8866	78.1307	7.1655	27.5146
	Standard Deviation	**0.3478**	24.2568	0.7314	258.6771	0.3510	26.1893
	Best	5.8069	0.0546	25.7383	0.0062	6.5208	0.6620
	Worst	7.3701	111.8152	27.9746	1.1219e+03	8.0564	89.6398
	Median	6.8667	5.6310	27.1346	0.6503	7.2043	20.8308
F6	Mean	6.5498e-10	6.0812e-10	0.7139	9.9607e-10	0.3435	3.6023e-31
	Standard Deviation	1.6644e-10	2.2654e-10	0.3548	6.0217e-10	0.1477	**9.8853e-31**
	Best	3.8422e-10	2.0192e-10	6.4081e-05	8.9491e-11	0.0597	0.0000
	Worst	1.0293e-09	1.0441e-09	1.2556	1.7876e-09	0.6379	4.4959e-30
	Median	6.3909e-10	5.5265e-10	0.6857	1.0657e-09	0.3523	8.6282e-32
F7	Mean	1.3569e-05	0.0046	0.0017	0.0466	0.0018	2.3283e-05
	Standard Deviation	**1.1634e-05**	0.0023	0.0015	0.0921	0.0014	2.1560e-05
	Best	7.1617e-07	0.0010	5.9667-04	2.3148e-04	1.6883e-04	1.2256e-06
	Worst	3.9854e-05	0.0087	0.0068	0.4063	0.0049	8.4515e-05
	Median	1.0738e-05	0.0043	0.0011	0.0088	0.0015	1.8608e-05
F8	Mean	-3.2454e+3	-2.8590e+3	-5.8693e+3	-1.6813e+3	-2.2116e+3	-3.692e+03
	Standard Deviation	420.1125	347.6819	1.0887e+3	175.9365	**139.7250**	186.6751
	Best	-3.9514e+3	-3.7161e+3	-7.9772e+3	-1.9765e+3	-2.4821e+3	-3.952e+02
	Worst	-2.6257e+3	-2.4041e+3	-2.7557e+3	-1.3797e+3	-1.9810e+3	-3.360e+03
	Median	-3.3198e+3	-2.9729e+3	-5.9411e+3	-1.7379e+3	-2.1900e+3	-3.656e+03
F9	Mean	**0**	11.1933	2.0587	5.7581	1.6298	8.3079
	Standard Deviation	**0**	5.3324	3.3974	4.5536	5.0518	2.8560
	Best	0	2.9849	1.1369e-13	0.9950	0	2.9849
	Worst	0	20.8941	11.1635	20.0345	18.3262	12.9345

F10	Mean	8.8818e−16	0.6157	1.0747e−13	0.3964	3.6419e−05	0.1906
	Standard Deviation	0	0.8981	**1.3880e−14**	0.8276	1.6287e−04	0.4700
	Best	8.8818e−16	5.2111e−06	7.9048e−14	1.8110e−05	8.8818e−16	4.4409e−15
	Worst	8.8818e−16	2.3168	1.2879e−13	2.3168	7.2839e−04	1.5017
	Median	8.8818e−16	1.0758e−05	1.0925e−13	1.4032e−04	4.4409e−15	7.9936e−15
F11	Mean	0	0.2750	0.0032	0.1143	0.0390	0.0191
	Standard Deviation	**0**	0.1323	0.0067	0.0520	0.0973	0.0257
	Best	0	0.1083	0	0.0246	0	0.0000
	Worst	0	0.5534	0.0215	0.2267	0.3520	0.0853
	Median	0	0.2325	0	0.1047	6.2339e−14	0.0099
F12	Mean	6.3064e−12	0.0563	0.0457	4.9053e−07	0.0777	9.3944e−11
	Standard Deviation	**2.1561e−12**	0.1419	0.0211	1.4601e−06	0.0424	2.1414e−10
	Best	2.6122e−12	2.0721e−12	0.0197	9.3515e−10	0.0167	1.5705e−11
	Worst	9.9885e−12	0.5026	0.0997	6.6404e−06	0.2022	8.1376e−10
	Median	6.7364e−12	9.3848e−12	0.0395	9.6065e−08	0.0698	1.5964e−11
F13	Mean	0.0011	0.0025	0.6470	0.0025	0.3145	0.0016
	Standard Deviation	**0.004**	1.6934e−11	0.2260	2.3636e−10	0.0786	0.0040
	Best	1.8693e−11	0.0110	0.3099	0.0110	0.1684	1.3498e−32
	Worst	0.0110	3.5422e−11	1.0122	1.0370e−07	0.4375	0.0110
	Median	2.8484e−11	0.9980	0.5954	0.9980	0.3208	1.6579e−32
F14	Mean	0.9980	0.9980	4.665	0.9980	1.0973	0.9980
	Standard Deviation	**1.8367e−16**	1.2478e−16	4.4838	2.9263e−16	0.4436	2.0587e−14
	Best	0.9980	0.9980	0.9980	0.9980	0.9980	0.9980
	Worst	0.9980	0.9980	12.6705	0.9980	2.9821	0.9980
	Median	0.9980	0.9980	2.4871	0.9980	0.9980	0.9980

(Continued)

TABLE 4.3 (Continued)
Comparison of the Proposed Algorithm with Other Optimization Algorithms at Benchmark Functions

Function		ASSA (Proposed)	SSA	GWO	GOA	SCA	HGAPSO
F15	Mean	4.0088e–4	8.2922e–04	0.0044	0.0067	0.0011	0.0003075
	Standard Deviation	**1.6608e–4**	3.4809e–04	0.0082	0.0092	3.4412e–04	0.0002
	Best	3.0749e–4	3.1456e–04	3.0785e–04	4.2627e–04	5.8031e–04	0.0003075
	Worst	7.7990e–4	0.0016	0.0204	0.0204	0.0015	0.0003075
	Median	3.0932e–4	7.5297e–04	3.9681e–04	7.8225e–04	0.0013	0.0003075
F16	Mean	–1.0316	–1.0316	–1.0316	–1.0316	–1.0316	–1.0316
	Standard Deviation	**6.4525e–15**	9.7833e–15	2.5449e–08	3.9422e–14	2.2270e–05	7.2568e–06
	Best	–1.0316	–1.0316	–1.0316	–1.0316	–1.0316	–1.0316
	Worst	–1.0316	–1.0316	–1.0316	–1.0316	–1.0315	–1.0316
	Median	–1.0316	–1.0316	–1.0316	–1.0316	–1.0316	–1.0316
F17	Mean	0.3979	0.3979	0.3979	0.3979	0.3993	0.3979
	Standard Deviation	**5.5549e–15**	4.7514e–12	7.2876e–05	3.7072e–12	0.0018	5.6245e–11
	Best	0.3937	0.3979	0.3979	0.3979	0.3979	0.3979
	Worst	0.3937	0.3979	0.3982	0.3937	0.4062	0.3979
	Median	0.3937	0.3979	0.3979	0.3937	0.3989	0.3979
F18	Mean	3.0000	3.0000	3.0000	3.0000	3.0000	3.0000
	Standard Deviation	**6.3699e–14**	9.7833e–14	2.5871e–05	3.4093e–13	1.7842e–05	9.3963e–12
	Best	3.0000	3.0000	3.0000	3.0000	3.0000	3.0000
	Worst	3.0000	3.0000	3.0001	3.0000	3.0001	3.0000
	Median	3.0000	3.0000	3.0000	3.0000	3.0000	3.0000

F19	Mean	-3.8624	-3.8628	-3.8617	-3.8241	-3.8546	-3.8627
	Standard Deviation	**1.8934e-14**	4.3318e-14	0.0021	0.1729	0.0021	6.5542e-02
	Best	-3.8628	-3.8628	-3.8628	-3.8628	-3.8604	-3.8628
	Worst	-3.8628	-3.8628	-3.8553	-3.0897	-3.8521	-3.0698
	Median	-3.8628	-3.8628	-3.8627	-3.8628	-3.8542	-3.8628
F20	Mean	-3.1955	-3.2324	-3.2772	-3.2681	-2.9750	-2.9964
	Standard Deviation	0.0443	0.0531	0.7073	0.0611	0.2468	**0.0273**
	Best	-3.3195	-3.3220	-3.3220	-3.3220	-3.1299	-3.0425
	Worst	-3.3195	-3.2007	-3.1381	-3.1999	-2.0468	-2.9810
	Median	-3.3195	-3.2030	-3.3220	-3.3220	-3.0133	-2.9810
F21	Mean	-10.1505	-8.5248	-9.3935	-7.2680	-2.7616	-10.1406
	Standard Deviation	**0.0012**	2.9578	1.8508	3.3745	1.9669	1.8780
	Best	-10.1524	-10.1532	-10.1529	-10.1532	-6.2621	-10.1532
	Worst	-10.1482	-2.6305	-5.0982	-2.6305	-0.4982	-2.6305
	Median	-10.1507	-10.1532	-10.1513	-10.1532	-2.6870	-10.1531
F22	Mean	-10.1340	-10.1392	-10.0193	-8.9924	-3.4448	-10.1392
	Standard Deviation	**1.1682**	1.1793	1.7073	2.9466	2.3686	2.2568
	Best	-10.4023	-10.4029	-10.4027	-10.4029	-6.8715	-6.4521
	Worst	-5.0860	-5.1288	-2.7658	-1.8376	-0.5211	-2.7658
	Median	-10.4001	-10.4029	-10.4013	-10.4029	-4.3716	-4.3715
F23	Mean	-9.9932	-9.1089	-9.7229	-8.5497	-4.1825	-9.1089
	Standard Deviation	**1.6646**	2.9668	2.4970	3.5320	2.4040	2.9678
	Best	-10.5359	-10.5364	-10.5362	-10.5364	-9.6145	-10.5364
	Worst	-5.1244	-2.8066	-2.4217	-2.4217	-0.9415	-2.6586
	Median	-10.5340	-10.5364	-10.5338	-10.5364	-4.6284	-10.5364

to be superior in 19 benchmark functions, including Sphere, Schwefel 2.22, Schwefel 1.2, Schwefel 2.21, Rosenbrock, Quartic, Rastrigin, Griewank, Penalized, Penalized 2, Foxholes, Kowalik, Six-hump camel back, Branin, Goldstein-price, Hartman 3, Shekel 5, Shekel 7, and Shekel 10, based on achieving the minimum standard deviation value. The SCA and GWO methods yielded better results for the Schwefel and Ackley functions, while HGPSO delivered the best performance on the Step and Hartman 6 benchmark functions. The findings highlight the overall effectiveness of ASSA in comparison to other optimization techniques.

The comparison of algorithms has been conducted on classical functions based on metrics such as mean, standard deviation, best, worst, and median over 20 independent runs. However, this approach does not assess individual runs, which raises the possibility that any observed superiority may be coincidental. Therefore, it is crucial to compare the results of each run to evaluate the significance of the outcomes. To determine the significance level for each run, the Wilcoxon rank sum statistical test was applied at 5% significance level, and the corresponding P-values for each benchmark are presented in Table 4.4. Strong evidence against the null hypothesis is provided by a P-value of less than 0.05, which implies that the superior final objective function values that the top algorithm achieved were not the result of chance. The best method for each test function is selected for statistical analysis, and it is contrasted with other algorithms one at a time. The smallest standard deviation is used to identify the optimal algorithm; if two algorithms have the same standard deviation, the method with the lowest mean value is deemed to be the best. The best algorithm in each function is marked with "N/A," which stands for "Not Applicable," because the best algorithm cannot be compared to itself.

As shown in the table, ASSA obtained the best results for 18 functions: specifically, F1, F3–F14, F16–F18, and F21–F23. In contrast, SCA and SSA achieved better results for functions F8 and F19, respectively, while HGPSO was identified as the top algorithm for F2, F15, and F20. According to the findings presented in Tables 4.3 and 4.4, ASSA consistently surpasses the other algorithms evaluated, highlighting the statistical significance of its superiority. In accordance with the No-Free Lunch (NFL) theorem [120], ASSA demonstrates a capacity to tackle problems that other algorithms struggle to solve efficiently.

4.2.5.2   The Need for Optimization of VMD Parameters

The selection of optimal parameters, such as the mode number and quadratic penalty factor (which controls frequency bandwidth), is crucial for determining VMD parameters, and these can be derived using the Ameliorated Salp Swarm Algorithm (ASSA) proposed in this study. By integrating opposition-based learning with a position updating concept into the basic SSA, the process is accelerated. The ASSA algorithm and the basic SSA have been compared. Figure 4.9 shows the convergence curves for both techniques. ASSA attains convergence more quickly than SSA, as seen in Figure 4.9.

In comparison to other methods, the accuracy of the suggested ASSA algorithm in optimizing VMD parameters has also been evaluated. Figure 4.10

TABLE 4.4

P-values Calculated for the Wilcoxon Rank Sum-test (Significance Level 0.05) Corresponding to the Results in Table 4.3

Function	ASSA	SSA	GWO	GOA	SCA	HGPSO
F1	N/A	6.7860×10^{-08}	6.7956×10^{-09}	6.7956×10^{-08}	6.7956×10^{-08}	6.7956×10^{-08}
F2	6.7956×10^{-08}	6.7956×10^{-08}	6.7956×10^{-08}	6.7956×10^{-08}	6.7956×10^{-08}	N/A
F3	N/A	6.7860×10^{-08}	6.7656×10^{-09}	6.7956×10^{-08}	6.7956×10^{-08}	6.7956×10^{-08}
F4	N/A	6.7478×10^{-08}	6.7956×10^{-08}	6.7956×10^{-08}	6.7956×10^{-08}	6.7956×10^{-08}
F5	N/A	0.0411	6.7956×10^{-08}	0.0962	0.0337	1.7936×10^{-04}
F6	N/A	0.0720	6.7956×10^{-08}	0.0810	6.9756×10^{-08}	6.776×10^{-08}
F7	N/A	6.7956×10^{-08}	6.7956×10^{-08}	6.7956×10^{-08}	6.9756×10^{-08}	0.1075
F8	6.7765×10^{-08}	1.2346×10^{-07}	6.7956×10^{-08}	6.5970×10^{-08}	N/A	6.5997×10^{-08}
F9	N/A	7.9043×10^{-09}	7.4517×10^{-09}	8.0065×10^{-09}	0.0096	7.8321×10^{-09}
F10	N/A	7.9334×10^{-09}	7.6187×10^{-09}	8.0065×10^{-09}	1.6310×10^{-07}	3.3187×10^{-09}
F11	N/A	8.0065×10^{-09}	0.0402	8.0065×10^{-09}	6.6826×10^{-05}	9.4038×10^{-06}
F12	N/A	0.7972	6.7956×10^{-08}	6.7956×10^{-08}	6.7956×10^{-08}	4.9511×10^{-08}
F13	N/A	0.0026	6.7956×10^{-08}	1.2493×10^{-05}	6.7860×10^{-08}	7.7336×10^{-05}
F14	N/A	N/A	6.4846×10^{-05}	N/A	3.5055×10^{-07}	N/A
F15	5.4753×10^{-05}	8.0065×10^{-09}	0.0055	1.5253×10^{-06}	1.0352×10^{-06}	N/A
F16	N/A	N/A	1.1129×10^{-07}	N/A	7.9919×10^{-09}	7.9919×10^{-09}
F17	N/A	N/A	8.0065×10^{-09}	N/A	8.0065×10^{-09}	N/A
F18	N/A	N/A	7.9919×10^{-09}	N/A	7.991×10^{-09}	N/A
F19	8.0065×10^{-09}	N/A	8.0065×10^{-09}	8.0065×10^{-09}	8.0065×10^{-09}	8.0065×10^{-09}
F20	2.8636×10^{-08}	2.7747×10^{-08}	3.0480×10^{-04}	2.8836×10^{-08}	6.7956×10^{-08}	N/A
F21	N/A	0.0057	0.2616	0.3637	6.7956×10^{-08}	0.3637
F22	N/A	2.7769×10^{-07}	0.0859	4.4162×10^{-05}	1.2346×10^{-07}	4.4162×10^{-05}
F23	N/A	8.3337×10^{-04}	0.9246	6.0054×10^{-04}	1.2330e-07	8.3337e-04

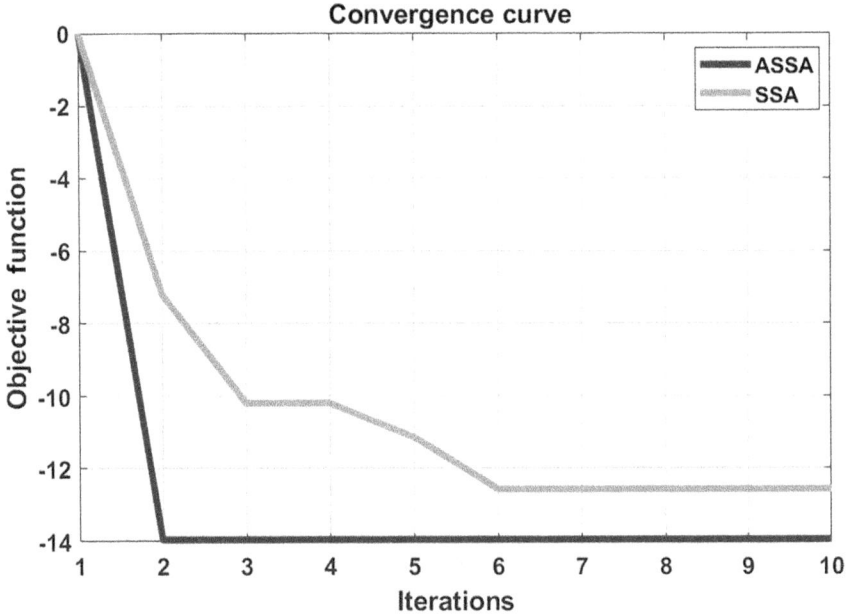

FIGURE 4.9 Convergence behavior for SSA and ASSA.

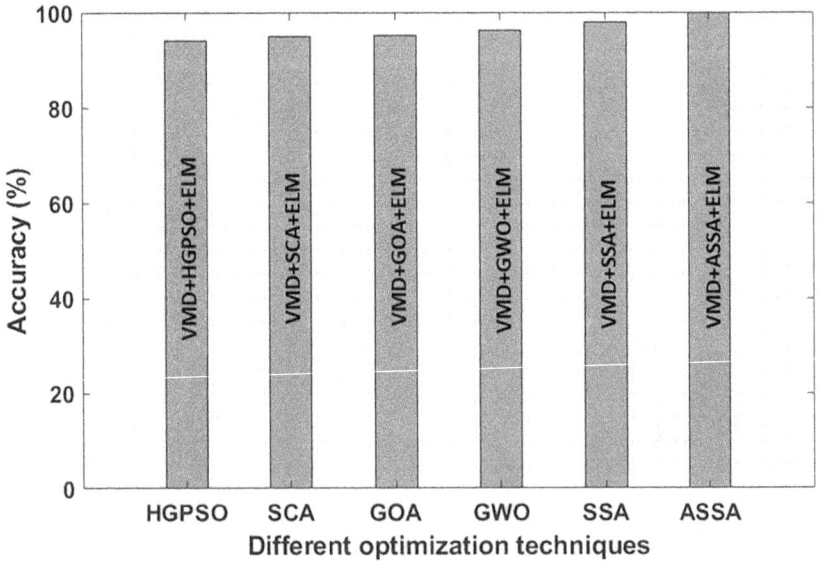

FIGURE 4.10 Comparison of various optimization methods regarding accuracy.

displays the results as bar charts. The figure makes it evident that ASSA performs better than the other optimization methods.

4.2.5.3 Results of the ELM Model and Its Comparison with Other Classification Models

The relevance of the extracted features is assessed using PCC. The descending values of the features' coefficients (weights), which are generated from this coefficient, are used to rank the features. Table 4.2 displays the feature weights, which were determined using PCC, and Table 4.5 provides a summary. Additionally, as shown in Figure 4.11, these weights are represented in bar graphs for comparison study. The feature "root mean square" (designated as Sl. No. 5) is the most significant of the 11 characteristics, as seen in Table 4.5 and Figure 4.11, since it has the largest weight, with the standard deviation coming in second.

The primary features found are used to generate a dataset. The ELM model then uses this dataset to categorize the various fault conditions. As suggested in [114], the ELM parameters are set up as follows: the kernel type is selected as radial basis function (RBF)-kernel, the kernel parameter is set to 0.01 and the regularization coefficient is set to 1. With a training time of 0.0012 seconds, the suggested method attains 100% training and 97.5% testing accuracy rates, respectively. Table 4.6 displays the results of comparing the ELM classifier's performance against alternative classification techniques. The outcomes show how well the suggested ELM-based method performs in identifying defect states from vibration data, clearly outperforming alternative classification methods.

TABLE 4.5
Weight of each Feature Obtained after Applying PCC

S. No.	Features	Weight of Features
1.	Standard Deviation $\left(x_{std}\right)$	0.2743
2.	Peak $\left(x_{p}\right)$	0.0552
3.	Skewness $\left(x_{ske}\right)$	0.0789
4.	Kurtosis $\left(x_{kur}\right)$	0.1435
5.	Root Mean Square $\left(x_{rms}\right)$	0.5223
6.	Peak Factor (PF)	0.0552
7.	Square Root Amplitude $\left(x_{sra}\right)$	0.0711
8.	Shape Factor (SF)	0.1619
9.	Impulse Factor (IF)	0.0259
10.	Wavelet Packet Decomposition (WPD) Energy	0.0915
11.	Permutation entropy	0.0850

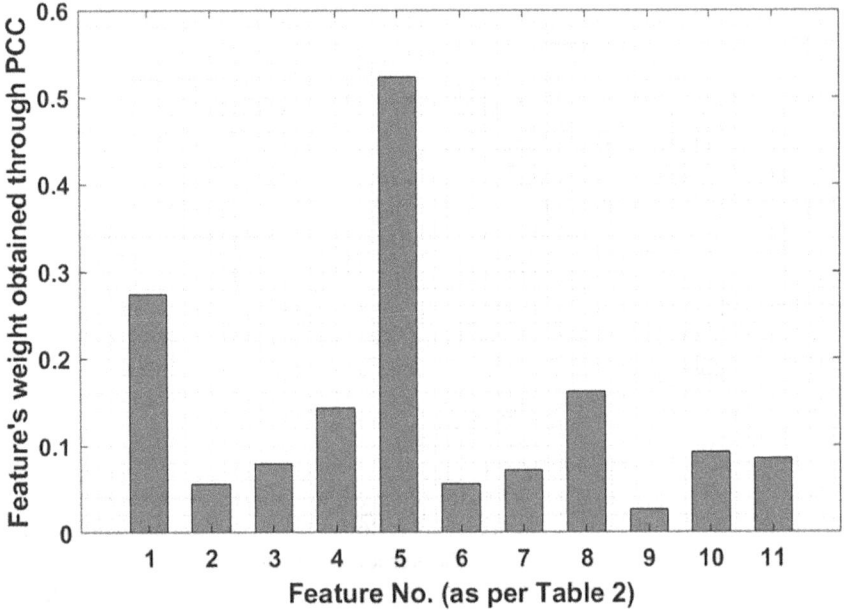

FIGURE 4.11   Weight of features obtained by PCC.

TABLE 4.6
Comparison of Performance of Different Classification Techniques with the Proposed Method Along with Training Time

S. No.	Classification Method	Training Accuracy		Training Time (sec) for One Iteration	
		With Ranking	Without Ranking	With Ranking	Without Ranking
1	KNN	85%	83%	19.06	23.58
2	SVM	87%	86.25%	25.01	27.45
3	Random Forest	85%	87%	18.56	26.5
4	Proposed method (ELM)	100%	97.5%	0.0012	0.0014

4.2.6 CONCLUSION OF CASE 1 STUDY

An opposition-based ASSA has been developed to enhance the adaptability of VMD for identifying impeller defects in centrifugal pumps. This algorithm adaptively selects the optimal combination of VMD parameters: mode number (K) and quadratic penalty factor α to align with the input signal. The key conclusions of the study are summarized as follows:

1. While the VMD parameters τ and ε have minimal effect on decomposition, K and α significantly influence results. Prespecifying K and α without prior signal knowledge is not recommended. Appropriate selection of the quadratic penalty factor α is crucial for noise reduction and bandwidth regulation.

2. VMD parameter optimization uses weighted kurtosis as its fitness function. This metric identifies sensitive modes and prevents information loss.

3. ASSA's performance was compared to other optimization algorithms across 23 benchmark functions (F1–F23), using mean, standard deviation, best, worst, and median values. ASSA achieved superior results (based on minimum standard deviation) on 19 functions (F1–F5, F6–F7, F9, F11–F19, and F21–F23). Wilcoxon testing further confirmed ASSA's statistically significant superiority on 18 functions (F1, F3–F7, F9–F14, F16–F18, and F21–F23).

4. Feature relevance was assessed using the PCC. Features were ranked by decreasing PCC value (weight). Root mean square was identified as the most prominent feature (highest weight) among the 11 features; standard deviation ranked second.

5. The developed ELM model achieved 100% training accuracy and 97.5% testing accuracy. Comparisons with other training methods showed superior performance in terms of both accuracy and computation time. While inherent, unstudied defects were included in the normal condition dataset, the results for the studied defect conditions were promising, highlighting a key advantage of this technique. The experiments demonstrate the method's capability for automatic centrifugal pump fault identification.

4.3 DIAGNOSIS OF BEARING DEFECTS IN CENTRIFUGAL PUMP (CASE 2)

Centrifugal pumps, capable of handling high fluid volumes, are susceptible to bearing defects caused by factors such as uneven forces, misalignment, insufficient lubrication, and manufacturing flaws. This work proposes a general normalized sparse filtering (GNSF)-based Wasserstein distance with maximum mean discrepancy (MMD) method for extracting bearing fault features from vibration signals. GNSF normalizes the feature matrix, while the Wasserstein distance with MMD performs fault clustering and highlights feature contributions.

4.3.1 THEORETICAL BACKGROUND

This section describes generalized sparse filtering, Wasserstein distance, MMD, and long short-term memory (LSTM).

4.3.1.1 Sparse Filter

Sparse filtering, a two-layer neural network for unsupervised feature learning [121], must satisfy three criteria: population sparsity, lifetime sparsity, and high dispersion (see Figure 4.12 for its basic configuration). The sparse filter

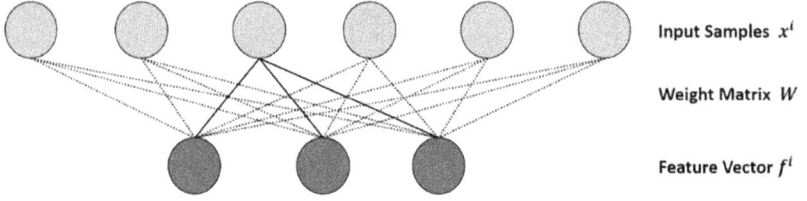

FIGURE 4.12   Schematic of generalized sparse filtering.

selectively activates a few features, differentiating features from different samples for improved sample discrimination and thus, ensuring distinct feature extraction from different samples.

In a sparse filter, the input consists of collected training samples $\left\{x^i\right\}_{i=1}^M$ of a signal, where $x^i \in \Re^{L \times 1}$ is a sample with N data points and M is the total number of samples. The output represents learned features $\left\{f^i\right\}_{i=1}^M$. The sparse filter learns the feature matrix $f^i \in \Re^{L \times 1}$ (with L learned features) using a weight matrix $W \in \Re^{L \times N}$. The mapping relationship is:

$$f_l^i = g\left(W_l x^i\right) \tag{4.27}$$

where f_l^i is the *lth* feature corresponding to the *ith* sample, W_l is the lth row vector of W, and $(.) = |.|$ is the absolute value function. Minimizing the cost function in a sparse filter induces generalized normalization, increasing competition among learned features and ensuring sparsity. As noted, f^i is a feature matrix whose rows are normalized using l_p-norms, resulting in:

$$\tilde{f_l} = \frac{f_l}{\left\|f_l\right\|_p} \tag{4.28}$$

where f_l is the *lth* row vector of f^i and $\|\cdot\|_p$ represents the l_p-norm. Subsequent normalization of each column of f^i using l_q-norms yields

$$\hat{f}^i = \frac{\tilde{f}^i}{\left\|\tilde{f}^i\right\|_q} \tag{4.29}$$

where $\tilde{f}^i$ is *ith* column vector of f^i and l_q-norms are denoted by $\|\cdot\|_p$.

Finally, the sparse filter minimizes cost function (C) as represented in Eq. 4.30 to optimize the weight matrix W using an l_r-norm penalty.

$$C = \text{sgn}\left(q - r\right) \cdot \sum_{i=1}^M \left\|\hat{f}^i\right\|_r \tag{4.30}$$

where $\text{sgn}(.)$ represents the sign function. Eq. 4.30 shows that if $q = r$, then $\left\| \hat{f}^i \right\|_r$ always equals 1, hindering cost function minimization. Standard sparse filtering [121] uses $p = 2$, $q = 2$, and $r = 1$. The non-smooth cost function is smoothed by incorporating the soft-absolute function (activation function), as shown in Eq. 4.31.

$$f = \sqrt{(Wx)^2 + \varepsilon} \tag{4.31}$$

where $\varepsilon = 10^{-8}$. The limited-memory Broyden–Fletcher–Goldfarb–Shanno (L-BFGS) algorithm [122, 123] minimizes the cost function C. The gradient of the cost function with respect to W is:

$$\frac{\partial C}{\partial W} = \left(\frac{\partial C}{\partial f} \cdot \frac{Wx}{\sqrt{(Wx)^2 + \varepsilon}} \right) . x^T \tag{4.32}$$

where

$$\frac{\partial C}{\partial f_l^i} = \frac{\partial C}{\partial \hat{f}_l^i} \cdot \frac{1}{\left(\sum_{i=1}^{M} \left(f_l^i \right)^p \right)^{1/p}} - \left(f_l^i \right)^{p-1} \cdot \left(\sum_{i=1}^{M} \left(\frac{\partial C}{\partial \hat{f}_l^i} \cdot f_l^i \right) \right) \cdot \frac{1}{\left(\sum_{i=1}^{M} \left(f_l^i \right)^p \right)^{1/p+1}} \tag{4.33}$$

$$\frac{\partial C}{\partial \tilde{f}_l^i} = \frac{\partial C}{\partial \hat{f}_l^i} \cdot \frac{1}{\left(\sum_{i=1}^{L} \left(\tilde{f}_l^i \right)^q \right)^{1/q}} - \left(\tilde{f}_l^i \right)^{q-1} \cdot \left(\sum_{i=1}^{L} \left(\frac{\partial C}{\partial \hat{f}_l^i} \cdot \tilde{f}_l^i \right) \right) \cdot \frac{1}{\left(\sum_{i=1}^{L} \left(\tilde{f}_l^i \right)^q \right)^{1/q+1}} \tag{4.34}$$

$$\frac{\partial C}{\partial \hat{f}_l^i} = \text{sgn}(q - r) . \left(\hat{f}_l^i \right)^{r-1} . \left(\sum_{i=1}^{L} \left(\hat{f}_l^i \right)^r \right)^{1/r-1} \tag{4.35}$$

4.3.1.2 Wasserstein Distance

The distance between features that were extracted via generalized sparse filtering is measured by the Wasserstein distance. The distance between feature vectors x and y in feature space M is denoted by $\rho(x, y)$. Because the Wasserstein distance converts one distribution into another during distance calculation, it is demonstrated to perform better than other probability measures such as Jensen-Shannon (JS) divergence and Kullback–Leibler (KL) divergence [124]. The Wasserstein distance between probability distributions P and Q is determined using Eq. (4.36).

$$W_p(P,Q) = \left(\inf_{\mu \in \Gamma(P,Q)} \int \rho(x,y)^z \, d\mu(x,y) \right)^{1/z} \tag{4.36}$$

where $\Gamma(P,Q)$ is the joint distribution of all the marginal distributions of P and Q in set $M * M$.

4.3.1.3   Maximum Mean Discrepancy (MMD)

MMD quantifies the similarity between two distributions. It operates on the principle that if two distributions generate samples such that the mean values of a function f (defined on the distributions' shared feature space) are equal for both distributions, then the distributions are considered identical. Given datasets $X^s = \left\{x_i^s\right\}_{i=1,\dots,n^s}$ and $X^t = \left\{x_j^t\right\}_{j=1,\dots,n^t}$ for distributions, P and Q, respectively, their MMD is expressed by Eq. 4.37.

$$MMD\left(X^s, X^t\right) = \left\| \frac{1}{n^s} \sum_{i=1}^{n^s} \varnothing\left(x_i^s\right) - \frac{1}{n^t} \sum_{i=1}^{n^t} \varnothing\left(x_j^t\right) \right\|_{\mathcal{H}} \tag{4.37}$$

where $\varnothing(.)$ is a nonlinear mapping function, which represents the mapping between two resembling distributions.

4.3.1.4   Long Short-Term Memory (LSTM) Network

LSTM, a type of recurrent neural network (RNN), is widely used and highly efficient [125]. As described by Hochreiter and Schmidhuber [125], LSTMs include input, output, and several control gates. The input gate data enables the network to predict the output. The LSTM processes sequential data step-by-step. Eqs. (4.38)–(4.43) detail the fundamental principles of LSTM operation.

$$c_*^t = \tanh\left(W_{xc}x^{t-1} + W_{hc}h^{t-1} + b_c\right) \tag{4.38}$$

$$i^t = \sigma\left(W_{xi}x^{t-1} + W_{hi}h^{t-1} + W_{ci}c^{t-1} + b_i\right) \tag{4.39}$$

$$f^t = \sigma\left(W_{xf}x^{t-1} + W_{hf}h^{t-1} + W_{cf}c^{t-1} + b_f\right) \tag{4.40}$$

$$c^t = f^t \odot c^{t-1} + i^t \odot c_*^t \tag{4.41}$$

$$O^t = \sigma\left(W_{xo}x^{t-1} + W_{ho}h^{t-1} + W_{co}c^{t-1} + b_o\right) \tag{4.42}$$

$$h^t = O^t \odot \tanh\left(c^t\right) \tag{4.43}$$

where x_i and h_t are the inputs of the memory cell. f^t, i^t, and O^t are the output of the control gate, input gate, and output gate, respectively. $W_{xc}, W_{hc}, W_{xi}, W_{hi}, W_{ci}, W_{xf}, W_{hf}, W_{cf}, W_{xo}, W_{ho}$, and W_{co} indicate the weight matrix. b_c, b_i, b_f, and b_o are the bias vectors. c^{t-1} is LSTM unit step value that can be obtained by recursive function h^{t-2} as shown in Eq. (4.43). Thus, LSTM's output can be defined as $f_y\left(h^t\right) = h^t W_y + b_y$.

4.3.2 Proposed Fault Diagnosis Approach Using GNSF Based on Wasserstein Distance with MMD

The proposed fault diagnosis scheme uses GNSF and LSTM as its two main learning stages. First, GNSF training optimizes the weight matrix W to extract features from the raw vibration signal. Then, the LSTM classifies the machinery's health conditions based on these learned features. A training dataset $\left\{x^i, y^i\right\}_{i=1}^M$ is created using M samples.

Here, $x^i \in \Re^{N \times 1}$ represents the *ith* sample with N data points, and y^i is its corresponding health condition label. The proposed fault diagnosis scheme proceeds as follows:

Step 1: Training

First, the GNSF model is trained with input and output dimensions N_{in} and N_{out}, respectively, using N_s overlapping segments (see Figure 4.13). The training set, comprising N_s segments $\left\{s^j\right\}_{j=1}^{MN_s}$, where $s^j \in \Re^{N_{in} \times 1}$ represents the *jth* segment with N_{in} data points, is then l_2 normalized, resulting in:

$$\tilde{s}^j = \frac{s^j}{\left\| s^j \right\|_2} \tag{4.44}$$

This normalization minimizes the negative impact of outliers, facilitating optimal solution finding. The normalized training set (S_n) then undergoes whitening to improve the sparse filter's generalization ability, using Eq. 4.45.

$$\mathrm{cov}\left(S_n\right) = EDE^T \tag{4.45}$$

where cov is the covariance matrix, E is the orthogonal matrix, and D is the diagonal matrix. The whitened training set (S_w) is represented by Eq. 4.46.

$$S_w = ED^{-\frac{1}{2}} E^T S_n \tag{4.46}$$

Finally, S_w trains the SF model, yielding an optimized weight matrix W that extracts discriminatory features from the training data.

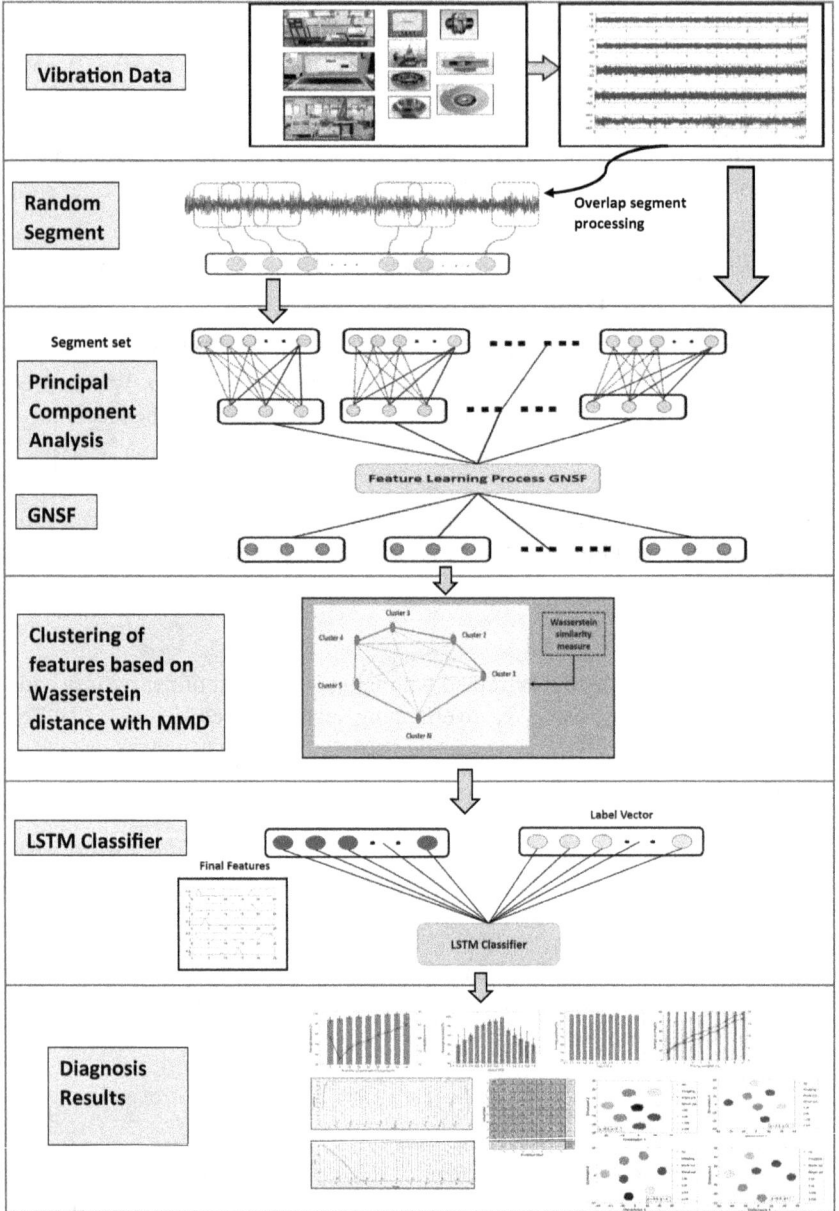

FIGURE 4.13   Architecture of the developed fault diagnosis method.

Step 2: Feature extraction

Post-training, each sample is divided into K segments $\left(K = \dfrac{N}{N_{in}}\right)$, creating a segment set $\{x_k^i\}_{k=1}^{K}$, where $x_k^i \in \mathfrak{R}^{N_{in} \times 1}$ is the kth segment of the ith sample. The SF maps each x_k^i to a local feature vector $f^i \in \mathfrak{R}^{N_{out} \times 1}$, producing a feature set $\{f_k^i\}_{k=1}^{K}$. The learned feature vector $f^i \in \mathfrak{R}^{N_{out} \times 1}$ is then represented by Eq. (4.47).

$$f_l^i = \frac{1}{K} \sum_{k=1}^{K} f_{k,l}^i \qquad (4.47)$$

where f_l^i denotes the lth learned feature of f^i, and $f_{k,l}^i$ is the lth local feature of f_k^i. These learned feature vectors f^i then form the feature matrix $f \in \mathfrak{R}^{N_{out} \times M}$.

Step 3: Clustering of the extracted features using the normalized GNSF parameters. The clustering of GNSF-extracted features (using normalized parameters) highlights individual feature contributions. This clustering uses a Wasserstein distance based on MMD; the Wasserstein distance measures interfeature distances, while MMD quantifies feature similarity.

Step 4: Using LSTM for fault diagnosis.

Finally, the LSTM classifies the different health states. Before LSTM training, the feature matrix columns are l_2 normalized. The trained LSTM model then diagnoses the health conditions of rotating machinery using test samples. Figure 4.13 presents a flowchart of the proposed algorithm. Figure 4.14 illustrates the designed unsupervised deep learning network incorporating the clustering method.

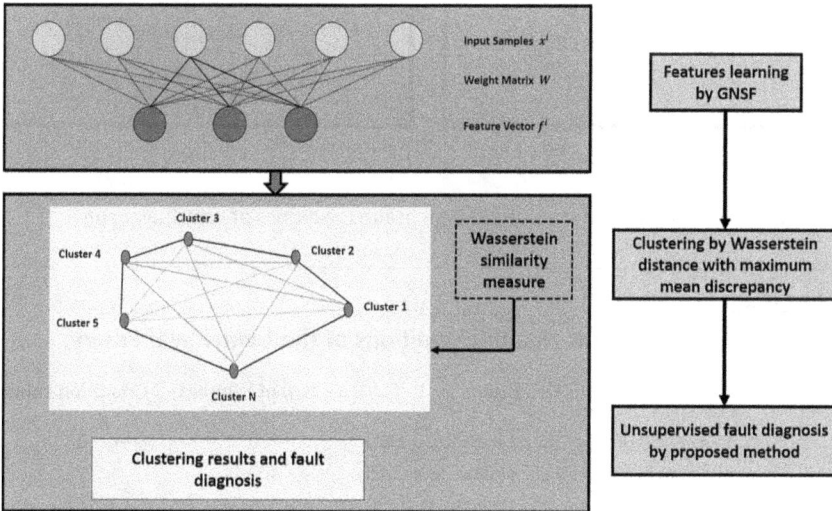

FIGURE 4.14   Schematic of the designed deep learning network (clustering).

4.3.3 Experimentation

4.3.3.1 Test Rig

The performance of the proposed method is validated using vibration data from a centrifugal pump test rig (refer to Figure 4.15 for a photograph of the pump; bearing details are provided in a previous section and in Figure 4.16; see Table 4.7).

FIGURE 4.15 Pictorial view of the centrifugal pump.

1 IR   1 OR

FIGURE 4.16 A pictorial view of different health conditions of centrifugal pump.

TABLE 4.7
Description of Different Health Conditions of the Centrifugal Pump

S. No.	Health Condition	No. of Samples	Condition Label
1	1 seeded hole of 1 mm dia. at inner race (1 IR)	400	0
2	2 seeded holes of 1 mm dia. at inner race (2 IR)	400	1
3	1 seeded hole of 1 mm dia. at outer race (1 OR)	400	2
4	2 seeded holes of 1 mm dia. at outer race (2 OR)	400	3

The study examines four bearing health conditions, with 400 samples collected for each condition. 10% of the data is randomly chosen for training the SF and LSTM models, while the remaining data is utilized for validation purposes. Each sample is divided into 200 training segments, each containing 700 data points, to prevent data leakage. PCA is performed, and the first c principal components (PCs) are used to construct the GNSF training matrix. Following the guidelines in [122], sparse filtering parameters are configured $(N_{in} = N_{out} = 100, N_s = 50)$, with 50 iterations of L-BFGS. To reduce variability, 20 experimental runs are performed; the performance and computation time are averaged across these runs, with standard deviations illustrated by error bars.

The number of PCs affects diagnostic performance and computation time by influencing input sample dimensionality and error. PCA was applied to each training sample's 200 segments to determine the optimal number of PCs (c). Diagnostic performance was analyzed for $p = 1, q = 1, r = 1$ at different c values (Figure 4.17; $c = 0$ indicates no PCA). PCA improves accuracy, consistency, and computational efficiency. Increasing c improves accuracy and reduces standard deviation but increases computation time; computation time becomes excessive for $c > 35$. A balance between accuracy and computation time is needed; experiments suggest an optimal c range of [15–35]. Therefore, c was selected from this range to optimize accuracy and computation time.

Prior sections explored the impact of normalization parameters. The method's performance was evaluated at various p values (with $q = 2, r = 2$). Figure 4.18 shows results for (a) when $p < q$, and (b) when $p > q$. High accuracy and

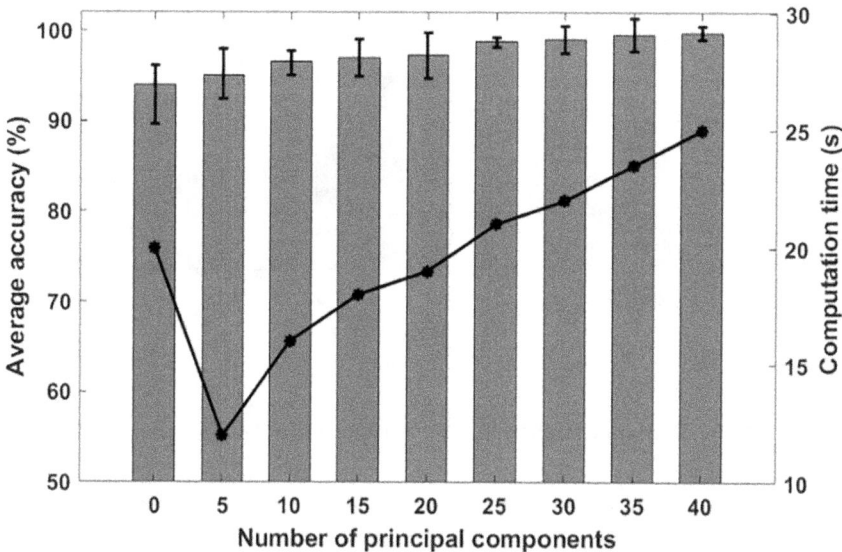

FIGURE 4.17 Results at different no. of PCs with $p = 1$ & $q = 2$.

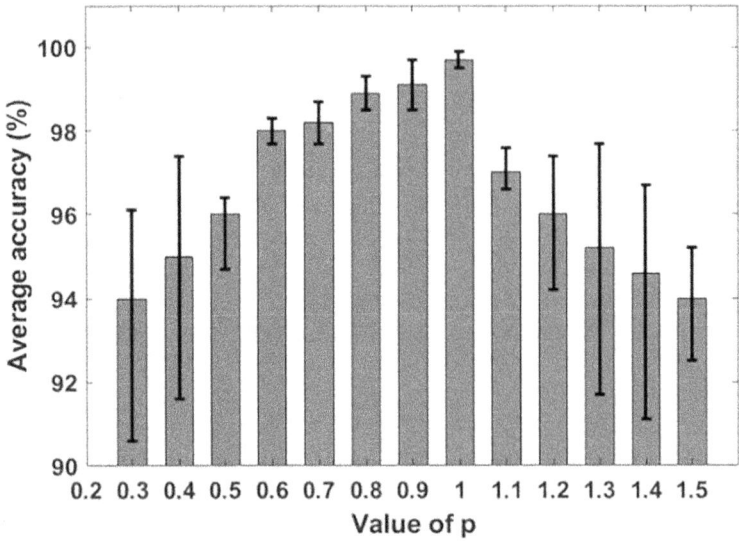

(a)

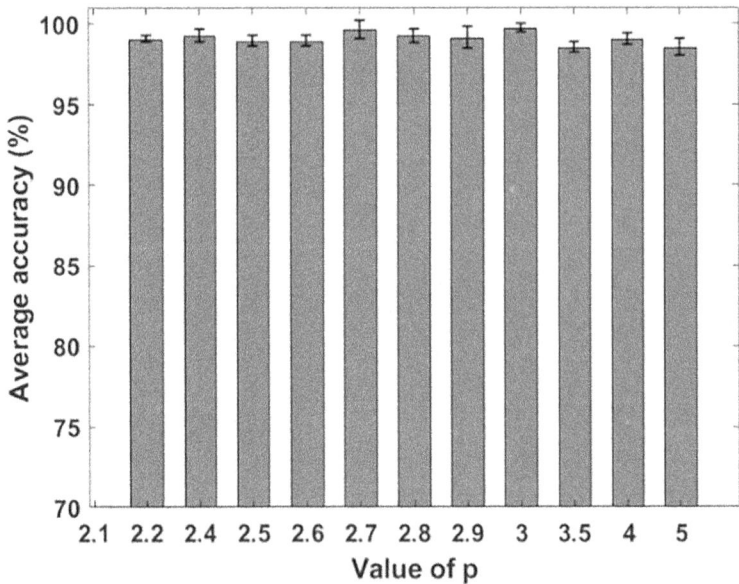

(b)

FIGURE 4.18   Diagnostic results different P values with $q = 2$ & $r = 2$ (a) $p < q$ (b) $p > q$.

consistent diagnostic results were observed for $0.5 < p < 1$ and $p > 2.5$; $p > 2.5$ yielded superior accuracy and lower standard deviation compared to $p < q$. Two key conclusions arise: (1) inappropriate p values negatively affect sparse filtering performance and (2) optimal accuracy with generalized sparse filtering is achieved when $2.5 \leq p \leq 3$ (with constant q and r).

This section analyzes the effect of varying q values on performance for (a) $p = 2.8$ and (b) $p = 0.8$. Sparse filtering failed to converge for both p values. Figure 4.19(a) shows that the method performs better when $p > q$. Figure 4.19(b) shows classification results for $p = 0.8$. Optimal q values fall within the range $1.5 < q < 2.5$; results are generally better for $q > 1.8$.

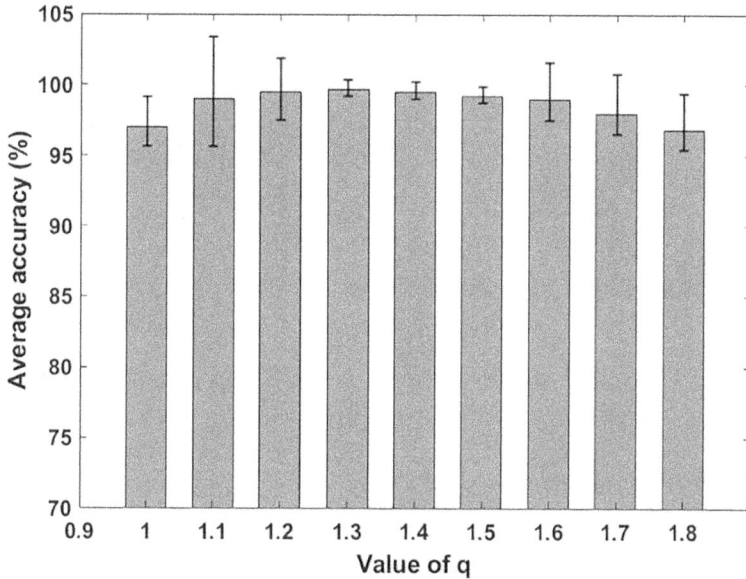

(a)

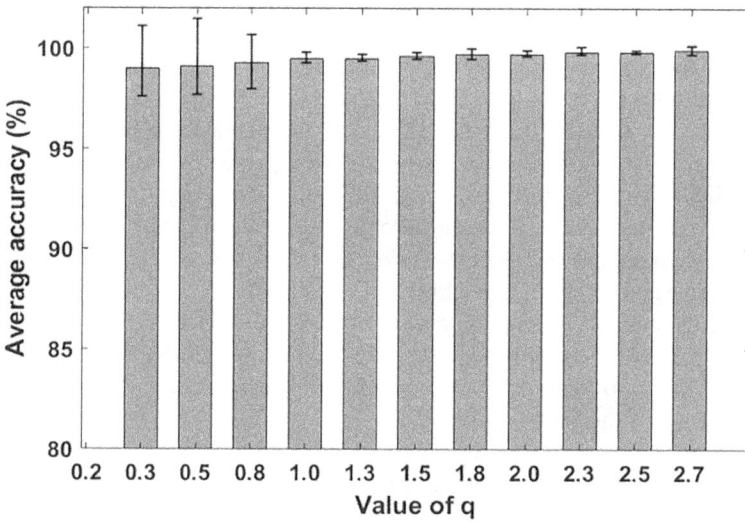

(b)

FIGURE 4.19 Diagnostic results at different values of q with $r = 2$ (a) $p = 0:8$ and (b) $p = 3$.

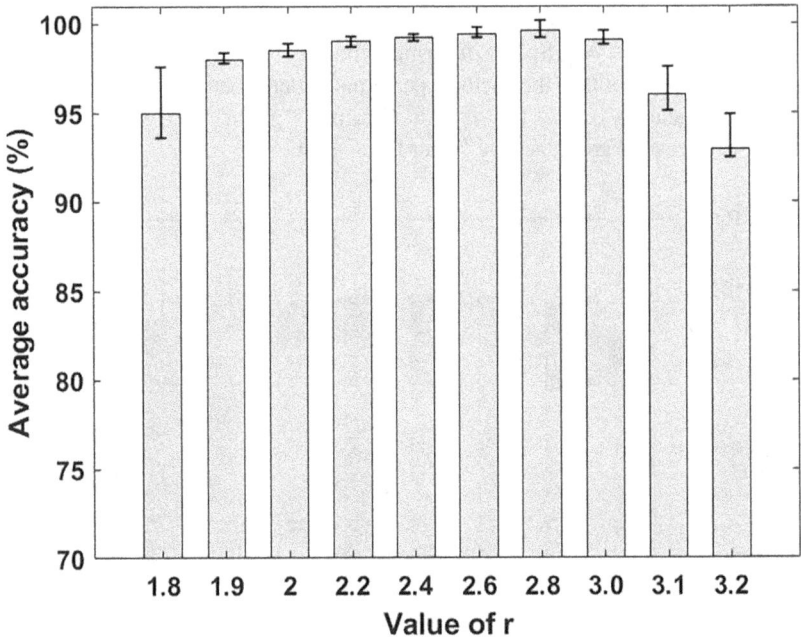

FIGURE 4.20 Diagnostic results at different values of r with $p = 3$ and $q = 2$.

Figure 4.20 shows the method's performance at various r values. Higher accuracy is achieved for $2 \leq r \leq 3$. r has minimal impact on diagnostic results. While row normalization extracts differential features and sparse filtering focuses on features related to the second derivative, extremely small or large r values negatively affect feature distinctiveness. Therefore, $r = 2$ is selected as the optimal value.

Results demonstrate the method's ability to accurately classify centrifugal pump health conditions for both $p > q$ and $p < q$. Optimal ranges for p and q ensure high efficiency and robustness. The study shows that generalized sparse filtering performs better (lower standard deviation) when $p > q$. The optimal row normalization parameter r is 2. p and q are interdependent; once p is chosen, q must be selected from an optimal (not too small or too large) range. This aligns with the theoretical approach. Figure 4.21 shows results from investigating the optimal range for the p/q ratio ($0.5 \leq p/q \leq 1.5$). The method performs best at a p/q ratio of 1.5, offering a wider range of normalization parameters while maintaining accuracy and stability.

Generalized sparse filtering extracts high-dimensional feature vectors. t-distributed stochastic neighbor embedding (t-SNE) [36] reduces these to two dimensions for visualization (Figure 4.22), illustrating why varying normalization parameters yield different accuracies. Figure 4.23 shows the confusion matrix $(p = 2.8, q = 2)$,

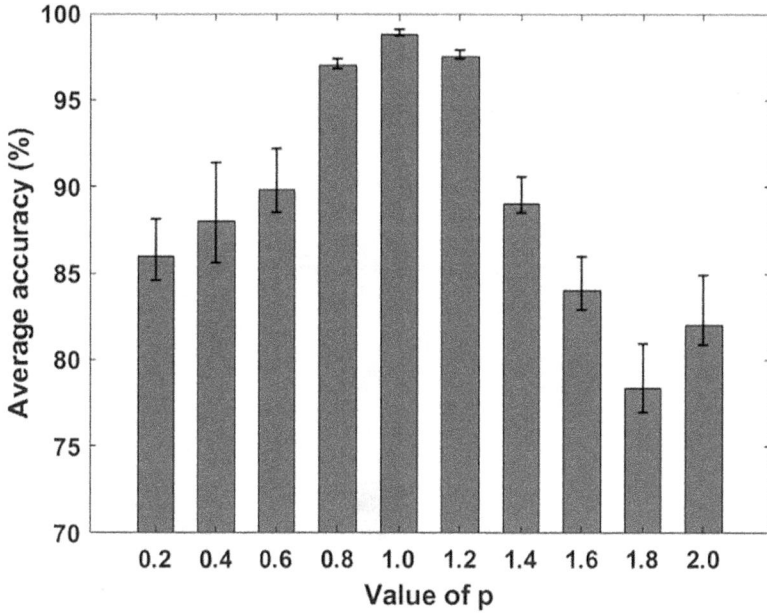

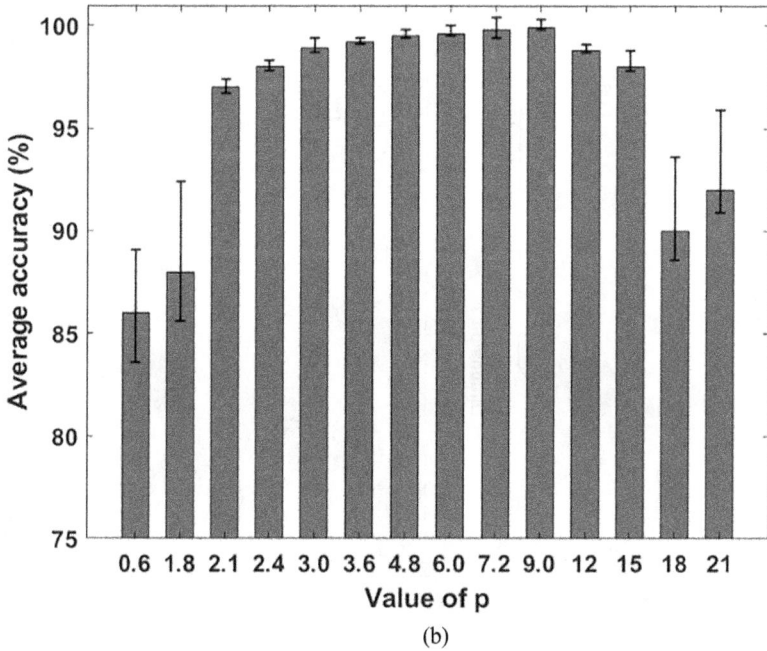

FIGURE 4.21 Diagnostic results with different values of p with $r = 2$ (a) $p / q = 0.5$ and (b) $p / q = 1.5$.

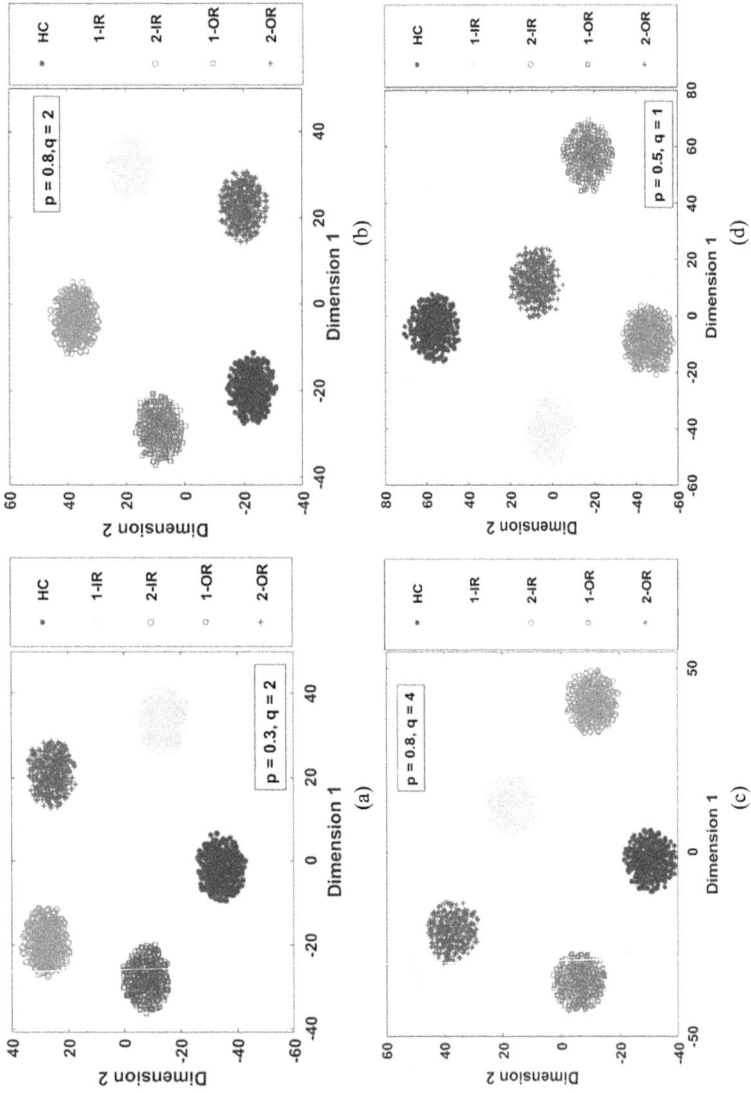

FIGURE 4.22 2D visuals of the features using t-SNE at 4 different conditions with $p < q$ and $r = 2$.

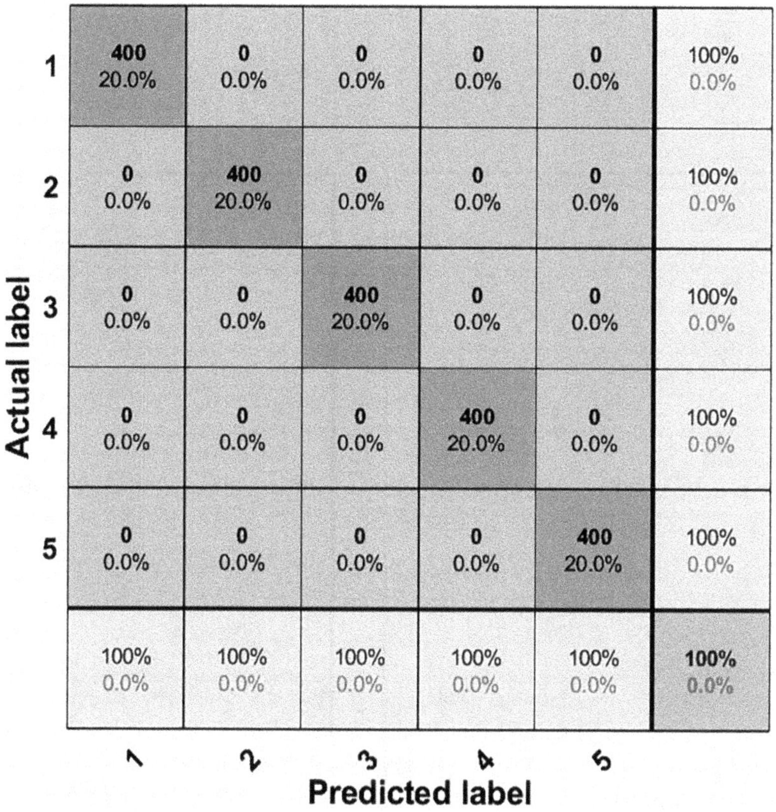

1	**400** 20.0%	0 0.0%	0 0.0%	0 0.0%	0 0.0%	100% 0.0%
2	0 0.0%	**400** 20.0%	0 0.0%	0 0.0%	0 0.0%	100% 0.0%
3	0 0.0%	0 0.0%	**400** 20.0%	0 0.0%	0 0.0%	100% 0.0%
4	0 0.0%	0 0.0%	0 0.0%	**400** 20.0%	0 0.0%	100% 0.0%
5	0 0.0%	0 0.0%	0 0.0%	0 0.0%	**400** 20.0%	100% 0.0%
	100% 0.0%	100% 0.0%	100% 0.0%	100% 0.0%	100% 0.0%	**100%** 0.0%

Predicted label

FIGURE 4.23 Confusion matrix at $p = 2.8$ and $q = 2$.

demonstrating 100% accuracy in classifying centrifugal pump health conditions. Figure 4.24 shows the LSTM classifier's accuracy at $p = 2.8$ and $q = 2$.

4.3.3.2 Comparison with Other Methods

Generalized sparse filtering's effectiveness was compared to standard sparse filtering using varying training sample sizes. Standard sparse filtering randomly extracts 200 overlapping segments $S \in \mathfrak{R}^{200 \times 200}$. per sample, then applies whitening $S_s \in \mathfrak{R}^{200 \times 400000}$. Features are learned using $p = 2.8, q = 2, r = 2$, and 20 PCs (from the range [15, 35]). Table 4.8 and Figure 4.25 present the results, accuracy increases, and standard deviation decreases with more training data. Standard sparse filtering shows higher computation time and standard deviation than the proposed method, even with larger training datasets. The proposed method achieves better performance with fewer training samples (e.g., exceeding the accuracy of standard sparse filtering at 10% of training data, even at only 1% of training data). At 35 PCs and 5% training data, the proposed method achieved 99.95% accuracy with 0.04% standard deviation, comparable to standard sparse

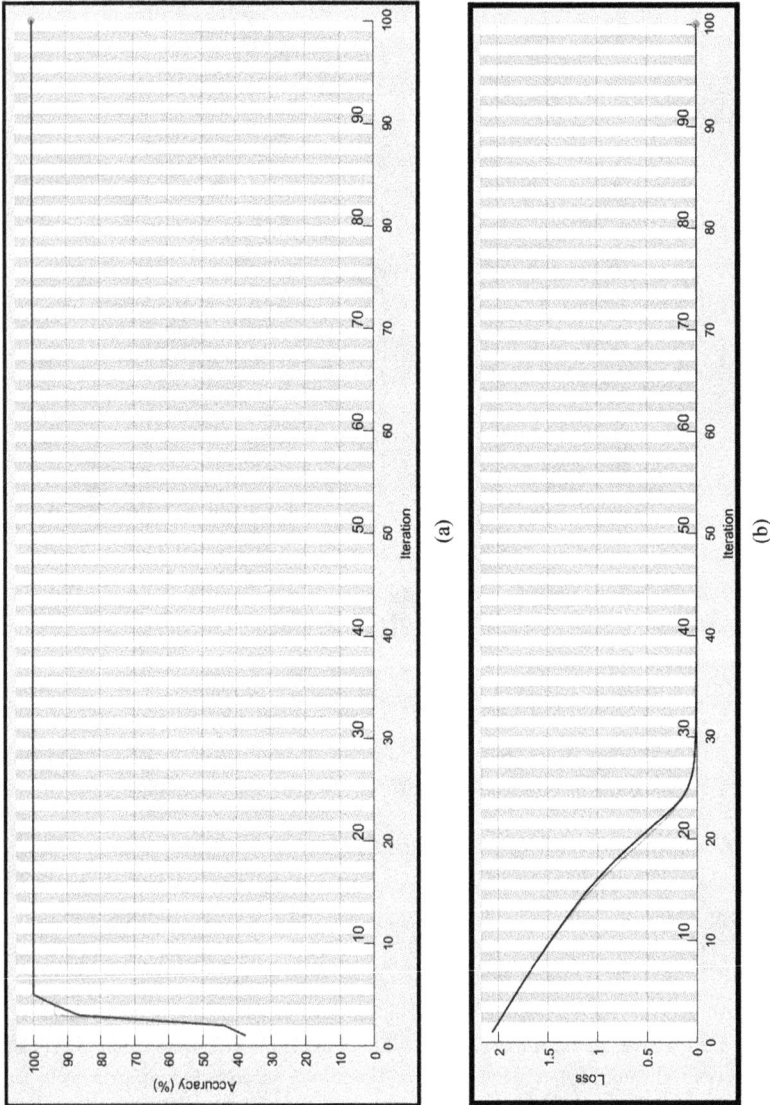

FIGURE 4.24 Training performance of LSTM model for centrifugal pump: (a) accuracy and (b) loss.

TABLE 4.8
Comparison of Different Sparse Filtering Methods for Centrifugal Pump

Methods	No. of Training Samples	No. Health States	Computational Time (s)	Standard Deviation (%)	Average Accuracy (%)
Standard sparse filtering	10 %	5	19.6	0.98	95.12
GNSF without PCA $(p = 2.8, q = 2)$	10 %	5	48.5	0.19	97.65
The proposed method $(p = 2.8, q = 2, 20$ PCs)	1 %	5	10.1	0.16	98.68
The proposed method $(p = 2.8, q = 2, 20$ PCs)	3 %	5	12.9	0.10	99.26
The proposed method $(p = 2.8, q = 2, 35$ PCs)	5 %	5	19.5	0.04	99.95

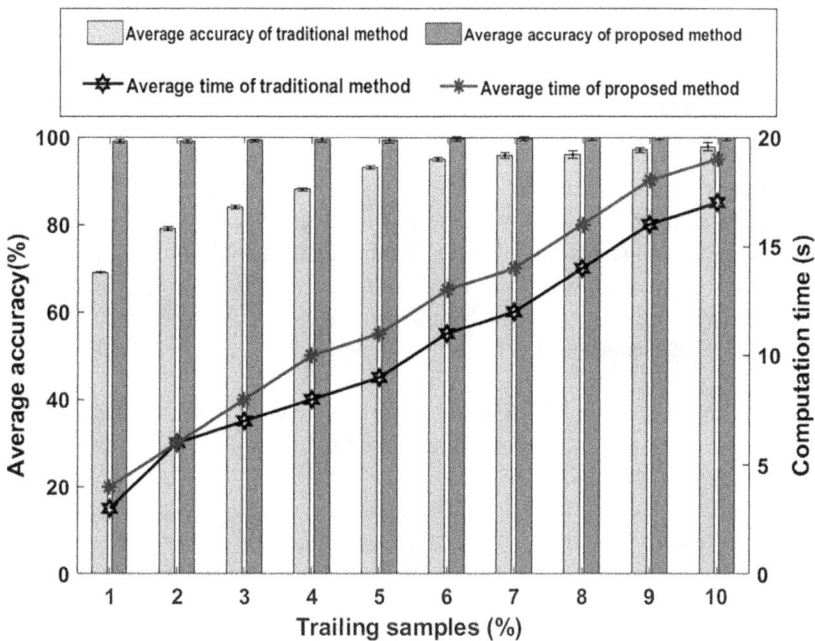

FIGURE 4.25 Comparison of diagnostic results under different methods with different number of samples.

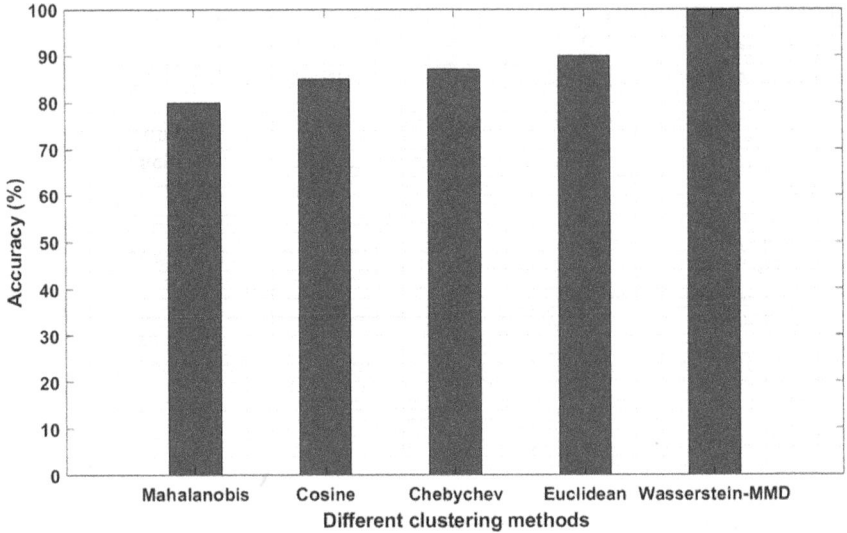

FIGURE 4.26 Comparison of different clustering methods in the fault diagnosis approach.

filtering's computational time. Comparisons with fault diagnosis methods using Mahalanobis, Cosine, Chebyshev, and Euclidean distances (Figure 4.26) demonstrate the proposed method's robustness and stability.

4.3.4 CONCLUSION OF CASE 2

This work proposes a novel unsupervised fault diagnosis method: GNSF combined with Wasserstein distance and MMD. The generalized $l_{r-p/q}$ norm objective function is optimized to improve feature sparsity and sparse filtering regularization. Wasserstein distance with MMD clusters features, highlighting their contributions. PCA preprocesses the data to remove correlations. An LSTM classifier identifies centrifugal pump faults. Results from the centrifugal pump dataset confirm the method's robustness. The study concludes that:

1. Optimization of sparse filtering parameters ensures the proposed method achieves more accurate and reliable results by adaptively extracting vibration signal features.
2. Feature clustering uses the Wasserstein distance with MMD, highlighting each feature's contribution to fault classification. Comparisons with traditional methods demonstrate the superiority of this new clustering approach.
3. The proposed method effectively identifies centrifugal pump health conditions even with limited training data.
4. The proposed method uses GNSF to extract discriminative features from the centrifugal pump. These features are then clustered using the Wasserstein distance with MMD, improving fault classification.

TABLE 4.9
Comparison of Fault Schemes of Cases 1 and 2

Methods	Impeller Defects		Bearing Defects	
	Average Accuracy (%)	Computational Time (s)	Average Accuracy (%)	Computational Time (s)
VMD-ASSA-ELM	100	12	100	13.8
GNSF ($p = 2.8, q = 2$, 35 PCs)	97.64	17.76	99.95	19.5

5. The proposed method accurately identifies various centrifugal pump health conditions, even with limited training data; for example, achieving 99.95% accuracy with only 5% of the samples used for training.
6. GNSF offers a wider range of normalization parameters, resulting in a more accurate and robust method than traditional sparse filtering.

4.4 COMPARISON OF METHODOLOGIES PROPOSED IN CASES 1 AND 2

Case 1's proposed method, while superior and robust for diagnosing centrifugal pump impeller defects, requires manual feature extraction. Case 2's method is fully automatic for bearing defect diagnosis. To compare them fairly, Case 1's method was applied to bearing defects, and Case 2's method to impeller defects. Table 4.9 shows the resulting accuracy and computation times.

Table 4.9 shows that Case 1's method achieves slightly higher efficiency and lower computation time compared to Case 2's method in fault diagnosis.

4.5 SUMMARY

This chapter investigates two centrifugal pump defect scenarios: impeller defects and combined impeller/bearing defects. For impeller defects, an ameliorated salp swarm algorithm (ASSA) optimizes VMD parameters using weighted kurtosis as the fitness function. The optimized VMD decomposes the signal into modes, and weighted kurtosis selects the most sensitive mode for feature extraction. PCC ranks features, indicating their importance and removing redundancy. These selected features train an ELM model to determine training and testing accuracy. For bearing defects, a GNSF method using Wasserstein distance with MMD is proposed. GNSF normalizes the feature matrix, and the Wasserstein distance with MMD performs fault clustering, highlighting feature contributions.

5 Fault Diagnosis of Bearing

5.1 INTRODUCTION

Rolling bearings are critical components found in numerous applications, such as wind turbines, helicopter gearboxes, aircraft engines, and high-speed trains, and they are susceptible to failures that can lead to costly downtime. Bearing defects generate cyclic impulses (repeating transients) in vibration signatures, which are valuable for condition monitoring. However, detecting these transients within complex signals presents difficulties due to noise and interference from other machinery components [126, 127]. Effective signal processing techniques are vital for early fault detection. Current methods designed to enhance fault characteristics can be categorized into three groups: resonance demodulation, decomposition, and time–frequency analysis [128–131].

A variety of signal processing techniques are available for detecting bearing faults, including wavelet decomposition [42], wavelet packet decomposition [43], linear mode decomposition (LMD) [132], empirical mode decomposition (EMD) [133], ensemble empirical mode decomposition (EEMD) [134], and complete ensemble empirical mode decomposition (CEEMD) [52]. These methods can effectively manage multi-frequency components. However, wavelet methods require pre-selection, which limits their adaptability. While both EMD and LMD are adaptive, they face challenges with mode mixing. EEMD and CEEMD help reduce mode mixing but can introduce difficult-to-remove noise [117]. Kumar and Kumar [134] provided a review of various vibration signal processing techniques for fault detection in rotary systems. Han et al. [133] utilized EMD, particle swarm optimization (PSO), and support vector machines (SVMs) for diagnosing gear faults under different loads. Buzzoni et al. [135] implemented automatic EMD for localized fault detection in multistage gearboxes. Variational mode decomposition (VMD) [136] offers enhanced bandwidth selection and noise suppression [137], but it requires predefined parameters [138]. Swarm decomposition (SWD) [139] addresses issues related to mode mixing and noise [140].

Contemporary researchers are increasingly relying on artificial intelligence for its capability to identify relationships within training patterns. Structures such as the Feed forward neural network (FFNN) [105], radial basis function neural network (RBFNN) [141], Elman neural network [142], and SVM [36] are various types of artificial neural networks (ANNs) employed to create classification models based on these relationships. Additionally, the extreme learning machine (ELM) is another approach utilized in classifying different machine components,

DOI: 10.1201/9781003614821-5

particularly in fault diagnosis and condition monitoring [114]. Abdoos [143] integrated ELM with VMD for predicting wind power, while Liu et al. [144] conducted numerical simulations to identify faults in gears and subsequently classified them using ELM.

Researchers have investigated various optimization techniques to determine the best ELM parameters, aiming to boost the model's accuracy. Finding the ideal parameter combination not only improves accuracy but also decreases computational time. Kang et al. [145] employed a genetic algorithm to optimize ELM parameters for defect identification in transformers. A restricted optimization-based method was presented by Shah et al. [146] to improve the ELM's gait detection training and testing accuracy. Furthermore, Yang et al. [147] found the ideal ELM parameters for diagnosing aero-engine faults by using quantum-behaved particle swarm optimization (QPSO).

An optimized ELM model for the automatic identification of bearing faults is presented in this work. SWD breaks down the raw signal into several modes. The mode with the lowest permutation entropy (PE) among these is thought to be the most significant or sensitive in terms of defect signatures. This prominent mode is used to pick features using a filter-based relief technique. To minimize data redundancy, characteristics are also ranked using score values obtained from statistical measurements. The ELM model is then trained using these chosen features as inputs. An opposition-based slime mold method is used to fine-tune the ELM parameters to attain optimal performance. An established process is used to validate the created ELM model.

5.2 PRELIMINARIES

5.2.1 Swarm Decomposition (SWD)

Apostolidis introduced an innovative method for signal decomposition called SWD [139]. By effectively parameterizing the approach, Swarm Filtering (SWF) can successfully isolate the key oscillatory component from the signal. The input signal $x[n]$ is viewed as the path taken by the prey in the swarm, and the processing is analogous to the swarm's hunting process, while the trajectory of the swarm corresponds to the output. A swarming model is developed to understand the theory behind SWF. Certain key concepts must be defined to construct this model. The position of the prey at the nth step is denoted as $P_{prey}[n]$. M refers to the number of swarms involved. $P_i[n]$ and $V_i[n]$ represent the position and velocity of the ith swarm at the nth step, respectively. Two distinct interactions govern the swarm's movement and hunting behaviors. The driving force $F_{dr-1}[n]$ represents the first interaction and is defined for the ith individual at the nth step as follows:

$$F_{dr-1}[n] = P_{prey}[n] - P_i[n-1] \qquad (5.1)$$

The cohesion force produces the second interaction and is characterized as an induced force acting on all members of the swarm, defined as follows:

$$F_{co-i} = \frac{1}{M-1} \sum_{i=1,i\neq j}^{M} f\left(P_i[n-1] - P_j[n-1]\right) \quad (5.2)$$

$$f(d) = -\text{sgn}(d)\ln\left(\frac{|d|}{d_c}\right) \quad (5.3)$$

Here, $\text{sgn}(.)$ and $\ln(.)$ denote the sign function and the logarithmic function, respectively. The function $f(.)$ can simultaneously apply the cohesion force in both attractive and repulsive manners. The variable d represents the distance between two swarm members, while $|.|$ indicates the absolute value. The critical distance is denoted as d_c, which typically regulates the swarm's distribution. Additionally, d_c also serves as the root mean square (RMS) of the input signal. To pursue the prey effectively, the swarm needs to update its position. Consequently, the velocity and position for the ith individual at the nth step are expressed as follows:

$$V_i = V_i[n-1] + \delta\left(F_{dr-1}[n]\right) + F_{co-i}[n] \quad (5.4)$$

$$P_i[n] = P_i[n-1] + \delta V_i[n] \quad (5.5)$$

In this case, d influences the adaptability of the swarm. The trajectory of the swarm, which represents the output of the SWF, is defined by the following Eq. (5.6).

$$y[n] = \beta \sum_{i=1}^{M} P_i[n] \quad (5.6)$$

The parameter β disrupts the order of M, with a smaller β value, such as 0.005, being favored to achieve a reasonable M. Both parameters d and M are crucial in governing the behavior of the swarm. The following criterion is used to determine the optimal values of these parameters.

$$\underset{\delta,M}{argmin} \sum_k \left\{|Y_{\delta,M}(k)| - |S(k)|\right\}^2 \quad (5.7)$$

In this context, $|Y_{\delta,M}(k)|$ and $|S(k)|$ denote the Fourier transforms of $Y_{\delta,M}[n]$ and $s[n]$, respectively. $Y_{\delta,M}[n]$ signifies the output of the SWF with parameters δ and M, while $s[n]$ represents the non-stationary multicomponent signal, which is combined with a mono-component. The primary goal of this process is to identify the values of δ and M. The SWF detects similarities in oscillatory components by comparing them with the non-stationary signal utilizing these parameters.

The association between swarm parameters and each individual frequency is referenced in [139].

$$M(\bar{w}) = \left[33.46\bar{w}^{-0735} - 29.1 \right] \tag{5.8}$$

$$\delta(\bar{w}) = -1.5\bar{w}^2 + 3.454\bar{w} - 0.001 \tag{5.9}$$

where $\bar{w}$ represents the normalized frequency. The value of M is established through a rounding operation.

The SWF is performed iteratively to identify the dominant oscillatory mode of the residue. The algorithm halts when the residual signal lacks any oscillatory mode. Additionally, SWF is concluded when the difference between two successive iterations falls below the threshold (T_th). The frequency band with the highest amplitude in spectral density is selected as a fitness function for optimization in each iteration. To enhance efficiency, the Savitzky-Golay (SG) filter is utilized, as it smooths the energy spectrum prior to identifying the highest peak [139, 140]. A predefined threshold for peak selection (PS_{th}) is established to minimize the search space. The optimal frequency w_m is determined using the following equations.

$$w_m = \underset{w}{argmax} (S'_{X_{it}}(w) > PS_{th}) \tag{5.10}$$

$$S'_{X_{it}} = SGfilter\left(S'_{X_{it}}(w)\right) \tag{5.11}$$

where S is the Fourier transform for the signal $x_{it}[n]$.

5.2.2 Permutation Entropy (PE)

For an arbitrary time-domain series $\{x(k), k = 1, 2, ..., N\}$, according to the embedding theorem, the delay embedding vector for D-dimensional data at time i is expressed in Eq. (5.12).

$$X_i^D = \left[x(i), x(i+\tau), x(i+2\tau), ..., x(i+(D-1)\tau) \right] \tag{5.12}$$

where D denotes the embedding dimension, which is greater than 2, τ represents the time lag, and i takes the values $1, 2, ..., N$. The notation $D!$ refers to the order of the symmetric group corresponding to the embedding dimension D, represented as S_D. This symmetric group encompasses all the permutations of length D [148].

Let $\pi_j = (j_1, j_2, ..., j_D) \varepsilon S_D$, where π_j denotes the symbol in S_D, π_j is permutation of X_i^D only when it becomes the unique symbol for S_D and satisfy the following two conditions:

$$x\left(i+\left(j_1-1\right)\tau\right) \leq x\left(i+\left(j_2-1\right)\tau\right) \leq \ldots \leq x\left(i+\left(j_D-1\right)\tau\right) \tag{5.13}$$

$$j_{s-1} < j_s \text{ if } x\left(i+\left(j_{s-1}-1\right)\tau\right) = x\left(i+\left(j_s-1\right)\tau\right) \tag{5.14}$$

Eq. (5.15) is used to obtain the relative frequency for each permutation π_j:

$$p\left(\pi_j\right) = \frac{\#\{x_i^D \text{ has type } \pi_j \mid 1 \leq j \leq N-\left(D-1\right)\tau\}}{N-\left(D-1\right)\tau} \tag{5.15}$$

where # denotes any constant number. As per Shannon's entropy of $D!$, the PE is defined as follows:

$$H_p\left(D\right) = -\sum_{\pi_j \epsilon S_D} p\left(\pi_j\right)\ln\left(p\left(\pi_j\right)\right) \tag{5.16}$$

The above equation is normalized by $\ln D!$ in the interval [0,1] and represented in the following manner:

$$H_p = \frac{H_p\left(D\right)}{\ln D!} = -\frac{1}{\ln D!}\sum_{i=1}^{k} p\left(\pi_j\right)\ln\left(p\left(\pi_j\right)\right) \tag{5.17}$$

The PE algorithm transforms the candidate time series into a symbolic series while maintaining the relationship between the current value and its equidistant past values [149]. To calculate PE, it is sufficient to understand the relationship between two sample points from the time series. This characteristic not only makes PE resistant to noise but also enhances its robustness. Additionally, PE measures the extent to which the time series deviates from randomness. A lower PE value indicates a more regular time series, meaning that an increase in the PE value corresponds to greater randomness in the time series [150]. Any change in PE amplifies variations in the time series.

5.2.3 EXTREME LEARNING MACHINE (ELM)

The ELM algorithm was introduced by Huang et al. [114]. It serves as a learning framework for single hidden layer feedforward neural networks (SLFN) and can be utilized for both classification and regression tasks. In ELM, the number of hidden layer nodes in the SLFN can be set adaptively during training. The input weights and biases of the hidden layer are chosen randomly, while the activation function depends on the specific problem [114, 151]. The weights linking the hidden layer to the output layer are computed analytically. Rather than randomly choosing the input weights and biases for the hidden layer, it is crucial to optimize both parameters to attain optimal fitness [145, 152].

5.2.3.1 ELM Model

For N arbitrary samples (x_i, t_i), where $x_i = [x_{i1}, x_{i2}, \ldots, x_{in}]^T \varepsilon R^n$ and $t_i = [t_{i1}, t_{i2}, \ldots, t_{im}]^T \varepsilon R^m$, a standard SLFNs with an activation function $f(x)$ and $\mathcal{N}$ neurons in the hidden layer can be mathematically modeled as

$$\sum_{i=1}^{\mathcal{N}} \beta_i f_i(x_j) = \sum_{i=1}^{\mathcal{N}} \beta_i f(a_i.x_j + b_i) = t_j; j = 1, 2, \ldots, N \tag{5.18}$$

where $a_i = [a_{i1}, a_{i2}, \ldots, a_{in}]^T$ represents the weight vector that connects the ith hidden node to the input nodes. b_i is the threshold associated with the ith hidden neuron. The weight vectors $\beta_i = [\beta_{i1}, \beta_{i2}, \ldots, \beta_{im}]^T$ are used to link the ith hidden neuron to the output neurons. The activation functions that can be chosen include "Sigmoid," "Sine," and "RFB."

Eq. (5.18) can be written as

$$H\beta = T \tag{5.19}$$

where

$$H(a_1, \ldots, a_{\mathcal{N}}, b_1, \ldots, b_{\mathcal{N}}, x_1, \ldots, x_{\mathcal{N}}) =$$
$$\begin{bmatrix} f(a_1.x_1 + b_1) & \cdots & f(a_{\mathcal{N}}.x_1 + b_{\mathcal{N}}) \\ \cdot & & \cdot \\ \cdot & \cdots & \cdot \\ \cdot & & \cdot \\ f(a_1.x_N + b_1) & \cdots & f(a_{\mathcal{N}}.x_N + b_{\mathcal{N}}) \end{bmatrix}_{N \times \mathcal{N}} \tag{5.20}$$

$$\beta = \begin{bmatrix} \beta_1^T \\ \cdot \\ \cdot \\ \beta_{\mathcal{N}}^T \end{bmatrix}_{\mathcal{N} \times m}, T = \begin{bmatrix} T_1^T \\ \cdot \\ \cdot \\ T_{\mathcal{N}}^T \end{bmatrix}_{\mathcal{N} \times m} \tag{5.21}$$

In this equation, H is the hidden layer output matrix of the neural network. The ith column of H is the ith hidden node output with regard to $x_1, x_2, \ldots, x_N$.

The conventional neural network learning algorithm requires the adjustment of multiple training parameters for the artificial network and often risks converging to a locally optimal solution. In contrast, the ELM algorithm removes the necessity for tuning the input weights and hidden biases of the network. The only requirement is to establish the number of nodes in the hidden layer. This approach yields a unique optimal solution, offering advantages such as rapid learning speed and improved generalization performance. As a result, training the SLFN is addressed as a linear equation using the least squares method (Figure 5.1).

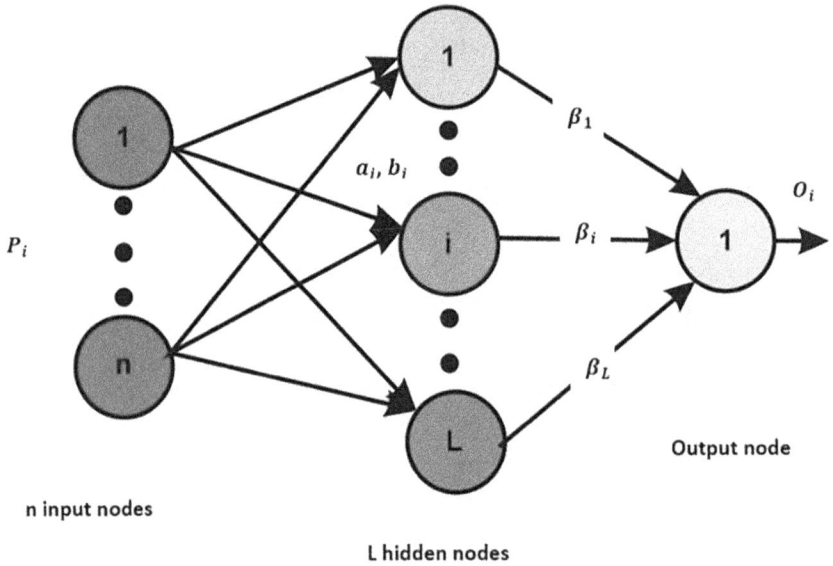

FIGURE 5.1   Extreme learning machine (ELM) architecture.

$$\left\| H\left(a_1,\ldots,a_N,b_1,\ldots,b_N\right)^\wedge \beta - T \right\| = \min_\beta \left\| H\left(a_1,\ldots,a_N,b_1,\ldots,b_N\right)\beta - T \right\| \qquad (5.22)$$

The least-square solution of linear Eq. (5.22) is given as

$$\beta = H^* T \qquad (5.23)$$

H^* indicates Moore–Penrose inverse of the hidden layer output matrix H.

5.2.4   SLIME MOULD ALGORITHM (SMA)

Li et al. [153] introduced a novel optimization algorithm known as the slime mould algorithm (SMA). In this context, "slime mould" refers to *Physarum polycephalum*, which is classified as a fungus. This eukaryotic organism thrives in cold climates, with its primary nutritional stage being Plasmodium. During this stage, the organic matter in the slime mould searches for food and secretes enzymes to aid in digestion. Drawing inspiration from the behavior of slime mould, Li et al. [153] developed a mathematical model. The slime mould navigates toward food sources by detecting their scent in the air. The following formula simulates this behavior of approaching food:

$$\overrightarrow{X(t+1)} = \begin{cases} \overrightarrow{X_b(t)} + \overrightarrow{vb}.\left(\overrightarrow{W}.\overrightarrow{X_A(t)} - \overrightarrow{X_B(t)}\right), & r < p \\ \\ \overrightarrow{vc}.\overrightarrow{X(t)}, & r \geq p \end{cases} \qquad (5.24)$$

where $\vec{X_b}$ represents an individual position with the strongest odor. $\vec{X}$ denotes the location of the slime. $\vec{X_A}$ and $\vec{X_B}$ are two randomly chosen individuals from the slime mould. $\vec{W}$ indicates the weight of the slime. The parameter $\vec{vb}$ is defined within the range of $[a,-a]$. The values for p, $\vec{vb}$, a, and $\vec{W}$ are specified in Equation (5.25).

$$p = \tanh|S(i) - DF| \tag{5.25}$$

Here, $1 \in 1,2,\ldots,n$, $S(i)$ is the fitness of $\vec{X}$. DF is the best fitness obtained in all iterations. As

$$\vec{vb} = [a,-a] \tag{5.26}$$

where a is defined as per Eq. (5.27).

$$a = arctanh\left(-\left(\frac{t}{max\_t}\right)+1\right) \tag{5.27}$$

The weight $\vec{W}$ is given as follows:

$$\frac{}{W(SmellIndex(i))} = \begin{cases} 1 + r.\log\left(\frac{bF - S(i)}{bF - wF} + 1\right), condition \\ \\ 1 - r.\log\left(\frac{bF - S(i)}{bF - wF} + 1\right), others \end{cases} \tag{5.28}$$

$$SmellIndex = sort(S) \tag{5.29}$$

where the *condition* indicates that the first half of the population is ranked according to $S(i)$. The variable r is a random value within the interval [0,1]. max_t refers to the maximum number of iterations. bF and wF represent the optimal fitness and the worst fitness in the current iterations, respectively. *SmellIndex* arranges the fitness values in ascending order. The following equation simulates the contraction of food by the slime mould.

$$\vec{X^*} = \begin{cases} rand.(UB - LB) + LB, rand < 1 \\ \vec{X_b}(t) + \vec{vb}.\left(W.\vec{X_A}(t) - \vec{X_B}(t)\right), r < p \\ \vec{vc}.\vec{X}(t), r \geq p \end{cases} \tag{5.30}$$

where UB and LB are upper and lower bound for the given search range, rand and r represent random values in the range [0,1].

5.2.5 OPPOSITION-BASED LEARNING

The global optimum is randomly selected to initiate any optimization algorithm, which in turn initializes individuals within a defined search space. Each individual updates their position based on their intelligence and behavior while searching for a solution. The computation time associated with these methods is influenced by the initial guesses. However, this time can be reduced by examining the opposite solution [154–156]. Subsequently, the solutions obtained from both the random choice and its opposite are evaluated to determine the best option. This optimal solution is used to initialize the individual, as verified by its fitness function. This approach not only lessens computational time but also enhances convergence speed. This technique is applied to each solution during the initialization process, which is conducted according to the following equations:

$$x_{ij} = x_j^{\min} + r_{ij}\left(x_j^{\max} - x_j^{\min}\right); \left(i = 1, 2, \ldots, NP; j = 1, 2, \ldots, D\right) \qquad (5.31)$$

$$x_{o_{ij}} = x_j^{\max} + x_j^{\min} - x_{ij} \qquad (5.32)$$

where x_{ij} denotes the initial population with an upper bound of $x_j^{\max}$ and a lower bound of $x_j^{\min}$. The term $x_{o_{ij}}$ represents the population derived from opposition-based learning. The variable r_ij is a uniformly distributed random number within the range of [0,1].

5.2.6 RELIEF-BASED ALGORITHM

The relief-based algorithm introduced by Kira et al. [157, 158] is based on instance-based learning principles. Relief calculates an intermediary statistic for each feature, which is utilized to gauge the relevance of the feature to the target. These statistics are referred to as feature weights (w[N]) and fall within a range from −1 (worst) to +1 (best). The code for the relief-based algorithm is presented in Figure 5.2.

Where the *diff* is defined for discrete and continuous features using equation (5.33) and (5.34), respectively.

$$diff\left(N, I_1, I_2\right) = \begin{cases} 0, if\ value\left(N, I_1\right) = value(N, I_2) \\ 1, otherwise \end{cases} \qquad (5.33)$$

$$diff\left(N, I_1, I_2\right) = \frac{\left|value\left(N, I_1\right) - value\left(N, I_2\right)\right|}{\max\left(N\right) - \min\left(N\right)} \qquad (5.34)$$

where $I_1 = R_i$ and I_2 is either $'H'$ or $'M'$ [159]

Input: for each training instance, a vector of the attribute value and predicted class.

$X \leftarrow$ *number of training instances*
$n \leftarrow$ *number of features*
$m \leftarrow$ *number of random training instances out of n used to update w*

Output: vector w of estimations of the weights of attributes.
- *Initialize all features weights $w[N] = 0.0$*
- *for $i = 1: m$*
- *Randomly select a target instance R_i*
- *Find a nearest hit 'H' and nearest miss 'M' (instances)*
  - *for $N = 1: n$*
  - $w[N] = w[N] - \dfrac{diff(N, R_i, H)}{m} + \dfrac{diff(N, R_i, M)}{m}$
  - *end for*
- *end for*

FIGURE 5.2 Pseudocode for relief-based algorithm.

5.3 PROPOSED ALGORITHM FOR FAULT IDENTIFICATION

The detailed procedure adopted for the automatic fault identification in the bearing is as follows:

- The obtained vibration signal is the input into the SWD with predetermined parameter ranges, which then decomposes it into various modes.
- The mode exhibiting the lowest PE is chosen as the dominant mode.
- The filter-based Relief algorithm is employed for selecting and ranking the features.
- With the extracted features, a dataset is created consisting of both training and test data.
- The training dataset is the input into the ELM, where its parameters (the weights connecting the input layer to the output layer and the biases in the hidden neurons) are optimized using the opposition-based SMA.
- The weight search range is established from 0.001 to 1000, while the biases have a range of 10 to 1000.
- Using the optimized ELM parameters, a classification model is developed for the purpose of fitness evaluation.
- The established ELM model is evaluated using the test dataset to determine the training and testing accuracy.

The whole procedure in the form of a flow chart is given in Figure 5.3.

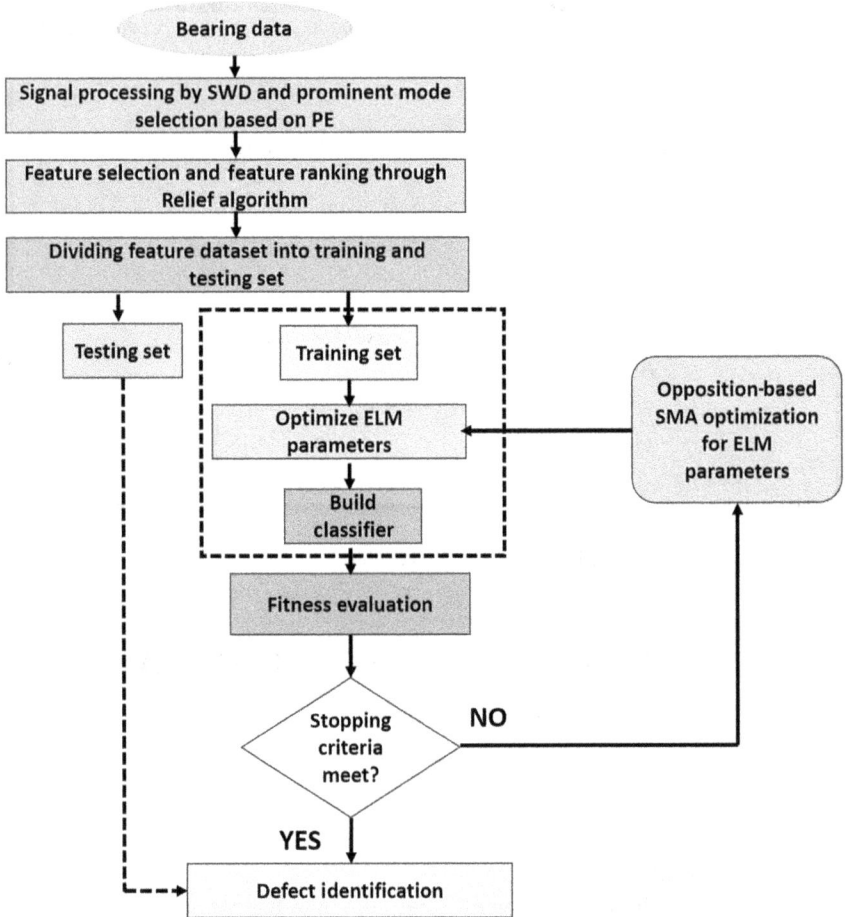

FIGURE 5.3 Flow chart for the proposed algorithm.

5.4 EXPERIMENTATION

5.4.1 EXPERIMENTAL SETUP AND DATA ACQUISITION

Experiments are conducted on a bearing setup, as shown in Figure 5.4. Vibration signals are captured from the bearing setup using a uniaxial accelerometer from PCB piezotronics with a sensitivity of 100 mV/g. The data acquisition system utilized for data collection is a 24-bit, 4-channel model from National Instruments. The accelerometer is secured to the bearing casing with wax and positioned perpendicular to the shaft's axis of rotation, enabling the capture of vertical acceleration at a sampling rate of 70 kHz. The taper roller bearing-2, designated as bearing number NBC 30205, is used to examine the seeded groove defect in (i) the outer race only, (ii) the inner race only, and (iii) a combination of both races,

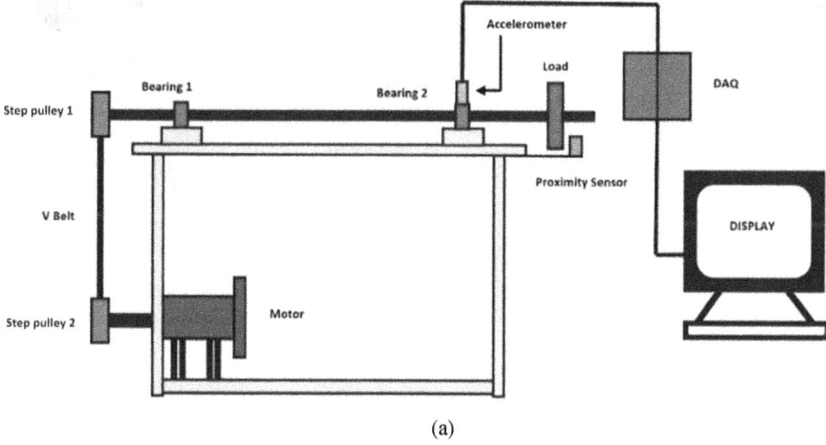

(a)

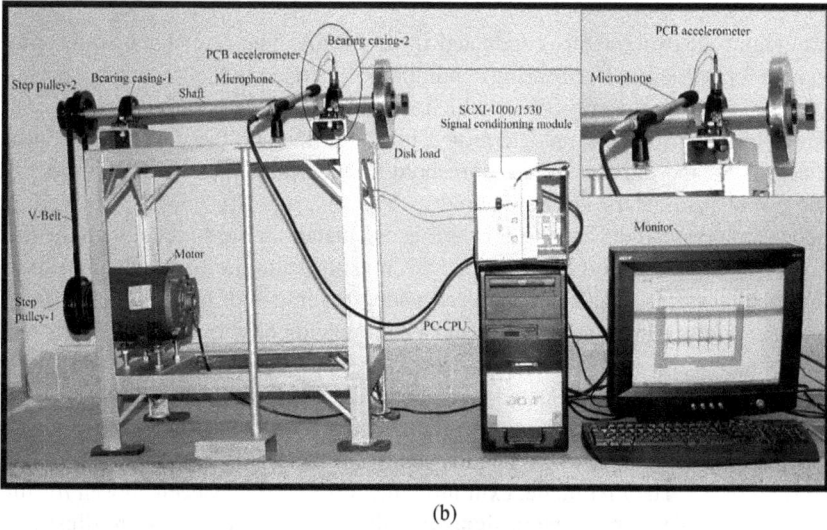

(b)

FIGURE 5.4   (a) Schematic of test rig. (b) Experimental test rig.

utilizing vibration signals. The groove widths for the outer and inner races are 0.5776 mm and 0.4714 mm, respectively, as illustrated in Figure 5.5. The analysis is performed for a signal length of 0.1 seconds, incorporating 7000 data points for each operating condition. The raw signal obtained from the bearing test rig is processed using the SWD method. PE is calculated for the decomposed modes generated by SWD, with the mode exhibiting the lowest entropy selected as the dominant mode.

Initially, the accelerometer gathers vibration data from the bearing test rig operating at a speed of 2050 rpm (equivalent to a frequency of 34.16 Hz) under healthy (defect-free) conditions. The test rig is maintained at a constant operating speed

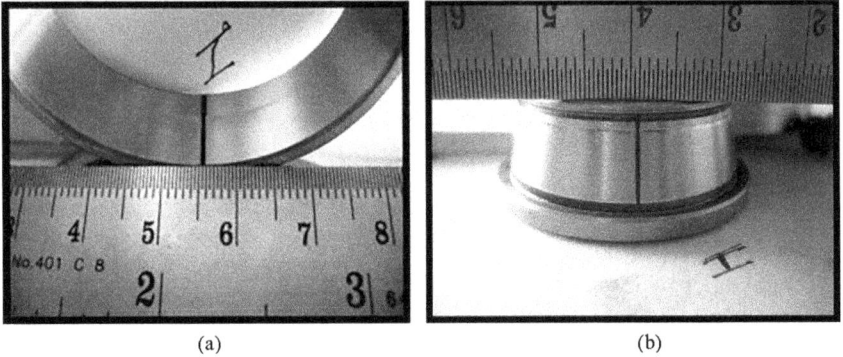

(a)								(b)

FIGURE 5.5 Different operating conditions: (a) outer race defect and (b) inner race defect.

throughout all experiments conducted in this study. The raw signal in the time domain for the defect-free bearing condition is illustrated in Figure 5.6(a). This raw signal is processed using SWD, which decomposes it into various modes. The parameters that must be predetermined for SWD include the threshold for peak selection P_{th} and the termination threshold StD_{th}, which halts the SWD process. The SWD parameters are configured as $P_{th} = 0.2$ and $StD_{th} = 0.2$. The various modes are depicted in Figure 5.6(b). PE is calculated for each mode, with the first mode yielding the lowest PE value of 1.69, making it the choice for feature extraction. A total of 50 signals are analyzed under conditions of healthy (defect-free), outer race defect, inner race defect, and a combination of both outer and inner race defects.

Similarly, the data is collected under conditions simulating an outer race defect and then decomposed into various modes using SWD, resulting in four modes with $P_{th} = 0.2$ and $StD_{th} = 0.2$. The PE values for these four modes are 1.96, 3.02, 4.21, and 4.32. The first mode, exhibiting the lowest PE, has been chosen for further analysis. The time-domain signal and the decomposed modes are illustrated in Figure 5.7(a) and (b), respectively.

The raw signal associated with the inner race defect condition is shown in Figure 5.8(a). With P_{th} set to 0.2 and StD_{th} at 0.2, the raw signal is decomposed into four modes, as illustrated in Figure 5.8(b). The PE values for the first, second, third, and fourth modes are 2.14, 2.79, 3.04, and 4.31, respectively. Since the first mode exhibits the lowest PE, it has been selected for further analysis.

The time-domain signal indicating the simultaneous occurrence of both outer-race and inner-race defect conditions is shown in Figure 5.9(a). Following the same procedure as in previous cases, the raw signal is decomposed into several modes using the SWD parameters set to $P_{th} = 0.2$ and $StD_{th} = 0.2$. The resulting four modes are illustrated in Figure 5.9(b). The PE values for the first, second, third, and fourth modes are 2.21, 3.01, 3.89, and 4.59, respectively. The mode with a PE of 2.21 has been identified as the dominant mode.

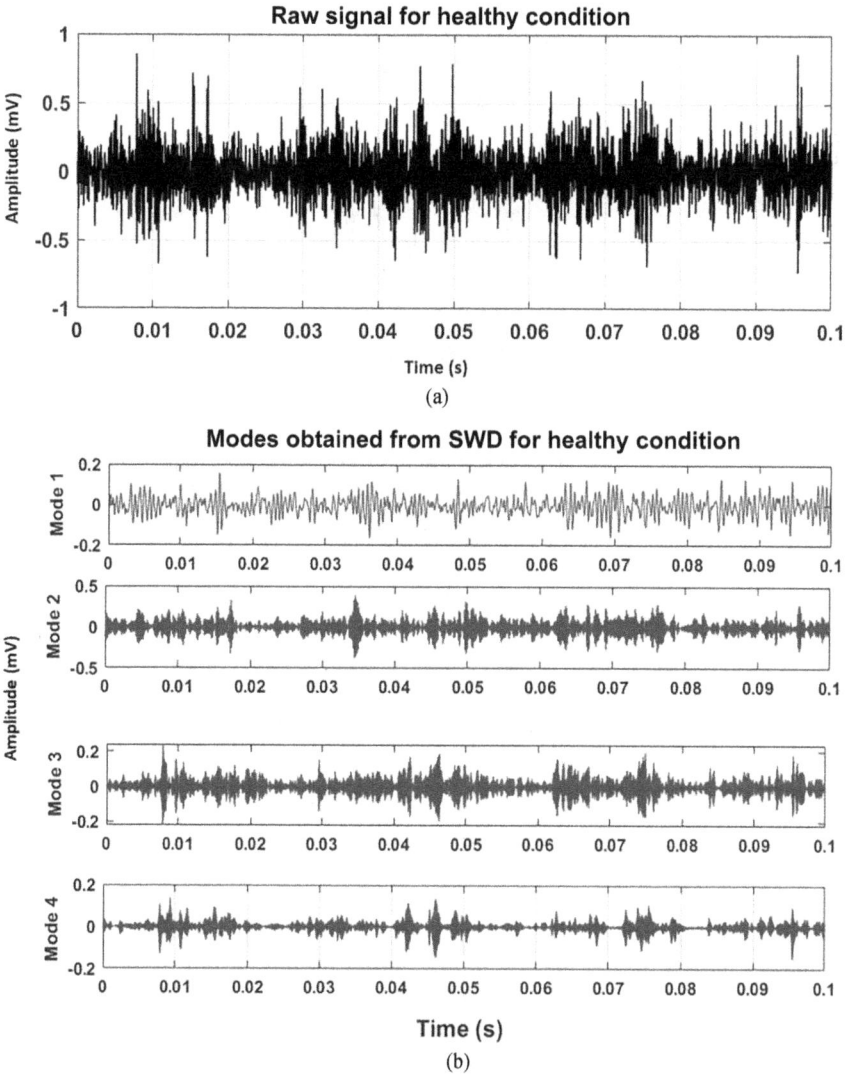

FIGURE 5.6 Vibration signal under healthy condition. (a) Raw signal and (b) different modes obtained through SWD.

5.4.2 FEATURE EXTRACTION

A total of 200 prominent modes of vibration signals are obtained from the SWD method, with 50 modes corresponding to each condition: healthy (defect-free), outer race defect, inner race defect, and a combination of outer and inner race defects, utilizing PE as the measurement index. The mode with the lowest PE value is regarded as the dominant mode for further analysis. Following this, 15 features

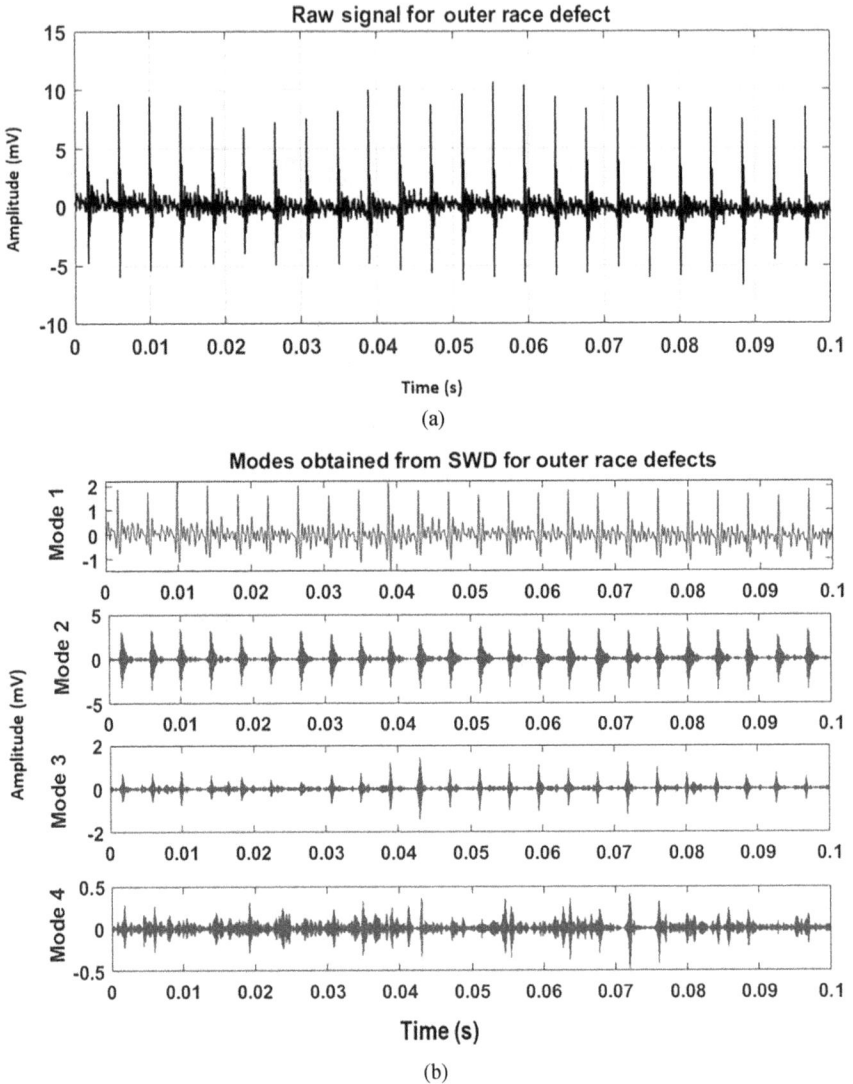

FIGURE 5.7 Vibration signal under outer race defect condition. (a) Raw signal and (b) different modes obtained through SWD.

are extracted from the prominent modes of the decomposed SWD. A list of these features along with their definitions is presented in Table 5.1. The extracted features are normalized within the range of [0,1] using the following mathematical expression:

$$Feat_{Normalized} = \frac{Feat - Feat_{min}}{Feat_{max} - Feat_{min}} \tag{5.35}$$

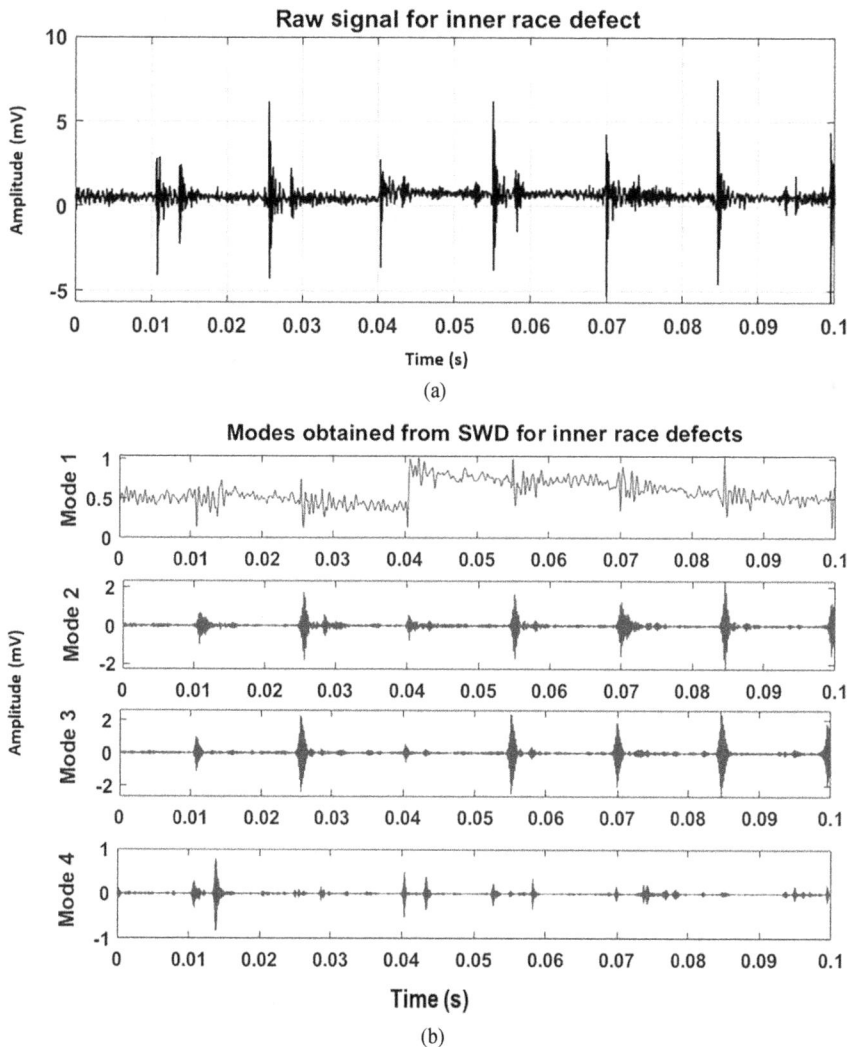

FIGURE 5.8 Vibration signal under inner race defect condition. (a) Raw signal and (b) different modes obtained through SWD.

where $Feat_{min}$ denotes the minimum value of a feature, while $Feat_{max}$ represents the maximum value of that feature.

In this context, x denotes the data, N represents the length of the data (i.e., the number of samples), and $x_k(n)$ indicates the decomposition coefficient for the kth sequence. The variable j corresponds to the level of wavelet packet decomposition (WPD) decomposition, while $s(k)$ signifies the spectrum of the signal x, and K represents the number of lines.

Fifteen features are extracted from the decomposed signal using SWD, covering both the time and frequency domains. To reduce data redundancy and

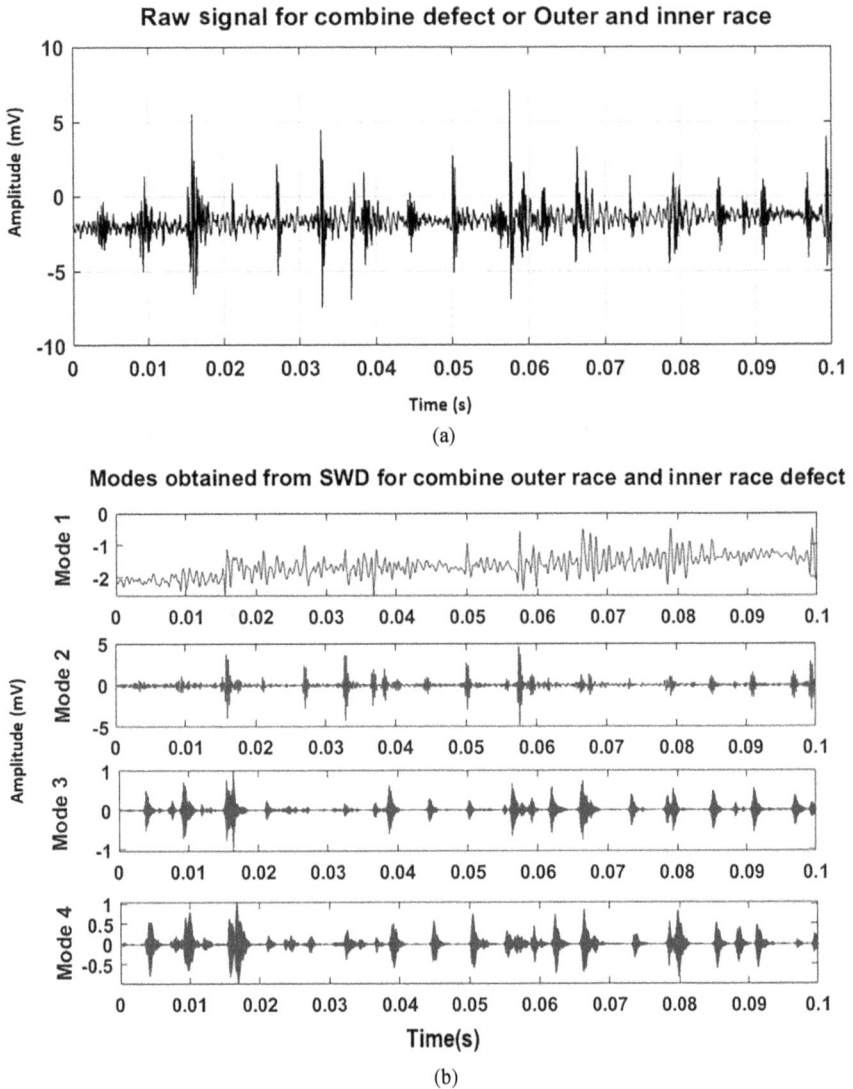

FIGURE 5.9 Vibration signal under combined outer race and inner race defect condition. (a) Raw signal and (b) different modes obtained through SWD.

determine which feature contributes the most, a filter-based feature selection method, that is, relief-based algorithm (RA) is utilized. This filter-based technique employs statistical measures to calculate a score for each feature, ranking them according to these scores. The scores for each feature are presented in Figure 5.10, while the ranks assigned to each feature are displayed in Figure 5.11. From these figures, it is clear that the RMS is the most important feature among the fifteen, as it holds the top rank, while the statistical parameter peak is in the second position.

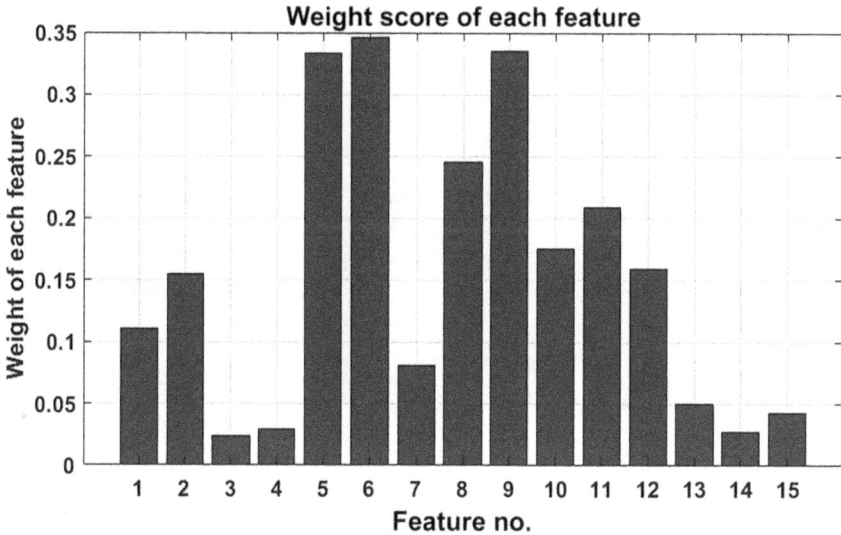

FIGURE 5.10   Weight (score) of each feature.

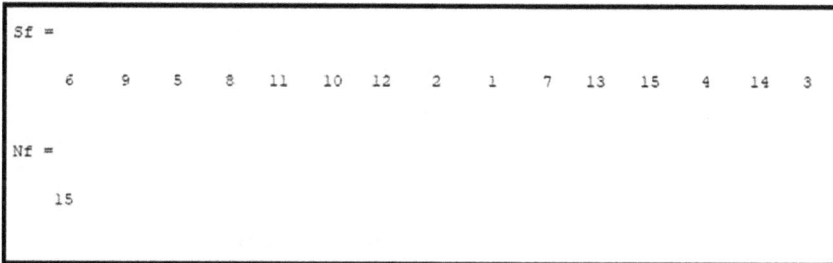

```
Sf =

    6    9    5    8    11   10   12    2    1    7    13   15    4    14    3

Nf =

    15
```

FIGURE 5.11   The rank assigned to features; Nf indicates the total number of features.

5.4.3 FITNESS EVALUATION

Based on the ranking of the prominent features, a dataset is constructed. This dataset is subsequently fed into the ELM, which classifies the different fault conditions. Within the ELM, two parameters need optimization: the input connection weight and the hidden biases of the single hidden layer feedforward network. These parameters are optimized using the opposition-based SMA. The optimized ELM classification method is then employed to compute the error, as expressed in Eq. (5.36).

$$f = \min(error) \tag{5.36}$$

where *error* is defined as $1 - Accuracy$, where *Accuracy* represents the proportion of samples that are correctly classified out of the total number of samples in the training set.

TABLE 5.1
Definition of Features

S. No.	Features Name	Features Definition
1	Wavelet Packet Decomposition (WPD) Energy	$WPD_i = \sum\limits_{n=1}^{N}\left|x_i(n)\right|^2 / \sum\limits_{k=0}^{2^j-1}\sum\limits_{n=1}^{N}\left|x_k(n)\right|^2$
2	Standard Deviation $\left(x_{std}\right)$	$x_{std} = \sqrt{\sum\limits_{i=1}^{N}\left(x(i)-x_m\right)^2 / N}$
3	Kurtosis $\left(x_{kur}\right)$	$x_{kur} = \sum\limits_{i=1}^{N}\left(x(i)-x_m\right)^4 / Nx_{std}^4$
4	Skewness $\left(x_{ske}\right)$	$x_{ske} = \sum\limits_{i=1}^{N}\left(x(i)-x_m\right)^3 / N$
5	Average $\left(x_{avg}\right)$	$x_{avg} = \dfrac{\sum\limits_{i=1}^{N}x(i)}{N}$
6	Root Mean Square $\left(x_{rms}\right)$	$x_{rms} = \sqrt{\sum\limits_{i=1}^{N}x(i)^2 / N}$
7	Variance $\left(x_{var}\right)$	$x_{var} = \dfrac{\sum\limits_{i=1}^{N}\left(x(i)-x_{avg}\right)^2}{N-1}$
8	Maximum $\left(x_{max}\right)$	$x_{max} = \max\left(x(i)\right)$
9	Peak $\left(x_p\right)$	$x_p = \max\left|x(i)\right|$
10	Peak Factor (PF)	$PF = {x_p}/{x_{rms}}$
11	Shape Factor (SF)	$SF = x_{rms} / \left(\sum\limits_{i=1}^{N}\left|x(i)\right| / N\right)$
12	Impulse Factor (IF)	$IF = x_p / \left(\sum\limits_{i=1}^{N}\left|x(i)\right| / N\right)$
13	Spectral Average	$MeanF = \dfrac{\sum\limits_{k=1}^{K}s(k)}{K}$
14	Spectral Variance	$RVF = \dfrac{\left(\sum\limits_{k=1}^{K}s(k)-MeanF\right)^2}{K-1}$
15	Spectral Kurtosis	$FK = \dfrac{\left(\sum\limits_{k=1}^{K}s(k)-MeanF\right)^4}{K.RVF}$

$$Accuracy = \frac{\sum_{i=1}^{N} \sum_{j=1}^{C} f(i,j) C(i,j)}{N} \tag{5.37}$$

where C denotes the labels of the dataset and N represents the number of samples in the training set. $C(i,j)$ is equal to 1 if the predicted class of sample i is j; otherwise, it is 0. The function $f(i,j)$ serves as a flag, indicating 1 if sample i belongs to class j.

The sensitivity, specificity, and precision are also assessed for identifying bearing defects using the optimized ELM. The mathematical expressions for sensitivity, specificity, and precision are provided in Equations (5.38), (5.39), and (5.40), respectively.

$$Sensitivity = \frac{TP}{(TP + FN)} \tag{5.38}$$

$$Specificity = \frac{TN}{(FP + TN)} \tag{5.39}$$

$$Precison = \frac{TP}{(TP + FP)} \tag{5.40}$$

where TP indicates true positive, TN is true negative, FP is false positive, and FN represents false negative.

5.5 RESULTS AND DISCUSSION

5.5.1 Purpose to Optimize ELM Parameters

ELM is a neural network designed for training models in classification and regression tasks. In ELM, the input weights (which connect the input layer to the output layer) and the biases of the hidden neurons are selected randomly. This randomness can lead to longer computation times for training the model and lower classification accuracy. Therefore, the optimal selection of these parameters is essential to address these issues and achieve an efficient ELM model. The outcomes of recognition using arbitrary values for input weights and biases are shown in Table 5.2.

Table 5.2 shows that the arbitrary selection of input weights and biases affects performance parameters (such as accuracy, sensitivity, specificity, and precision) and leads to suboptimal outcomes. As a result, the opposition-based SMA is utilized to optimize the parameters of the ELM. For this application, a population size of 30 is employed, with the maximum number of iterations set to 10, serving as the stopping criterion for this problem. The optimal classification error during training is achieved as zero in the second iteration, as shown in Figure 5.12, indicating a training accuracy of 100%. The time taken by the opposition-based SMA

TABLE 5.2
Effectiveness of ELM for Arbitrary Values of Input Weight and Biases

S. No.	Input Weight	Biases	Accuracy (%)		Sensitivity (%)		Specificity (%)		Precision(%)	
			Training	Testing	Training	Testing	Training	Testing	Training	Testing
1	3.5624	159.7268	92.67%	89.04%	90.92%	92.02%	85.58%	88.82%	88.71%	89.27%
2	4.9283	456.2948	85.65%	78.72%	85.25%	80.56%	80.98%	87.65%	83.59%	87.54%
3	11.6578	358.78	90.85%	80.29%	80.59%	87.15%	82.76%	85.29%	86.82%	82.47%
4	2.8954	571.1072	88.60%	85.75%	79.35%	88.52%	81.59%	80.62%	90.39%	88.82%

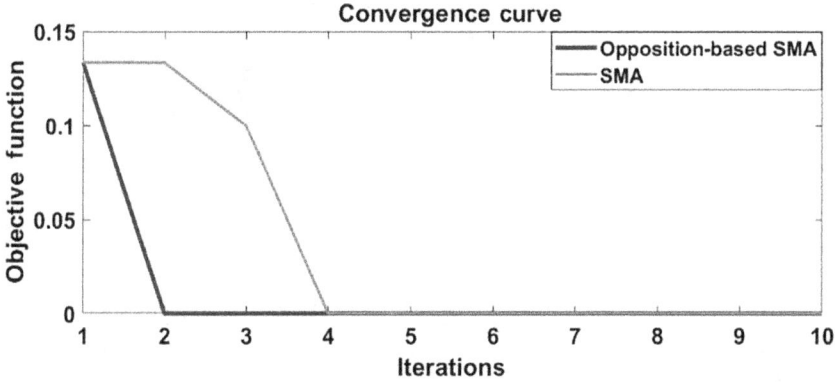

FIGURE 5.12   Convergence behavior for simple SMA and opposition-based SMA.

to optimize the ELM parameters is 0.0023 seconds. The optimized values for the input weights and biases in the hidden layers are determined to be 0.0789 and 500, respectively. With this combination of input weights and biases, the sensitivity, specificity, and precision are calculated to be 100%, 98.95%, and 100%, respectively. Additionally, a confusion matrix is constructed at this input weight (0.0789) and biases (500) in the hidden layers of ELM, reflecting various classes of bearing faults, including those with only outer race defects, only inner race defects, and both outer and inner race defects concurrently. The confusion matrix, presented in Figure 5.13, further validates the robustness of the proposed identification approach.

5.5.2   RESULTS OF THE FAULT IDENTIFICATION SCHEME AND ITS COMPARISON WITH OTHER CLASSIFIERS

The results from the ELM model are compared to those obtained from k-nearest neighbors (KNN), SVM, decision tree, and random forest classification models. The classification is performed for features both with and without ranking, and these findings are summarized in Table 5.3. As shown in Table 5.3, ELM surpasses other classifiers, achieving an accuracy of 100%. In contrast, KNN achieves an accuracy of 77.50%, SVM reaches 87.50%, the decision tree computes 83%, and the random forest achieves an accuracy of 86% for the provided data.

The accuracy of the proposed opposition-based SMA algorithm combined with the ELM classifier in identifying faults for each defect case is presented in Figure 5.14. A comparison with various classifiers (KNN, SVM, decision tree, and random forest) is also included in the same figure. The notations ND, OR, IR, and OR&IR in the figure represent no defect, outer race defect, inner race defect, and a combination of both outer and inner race defects, respectively. It is evident from this representation that the ELM classifier achieves the best performance when integrated with the proposed opposition-based SMA algorithm

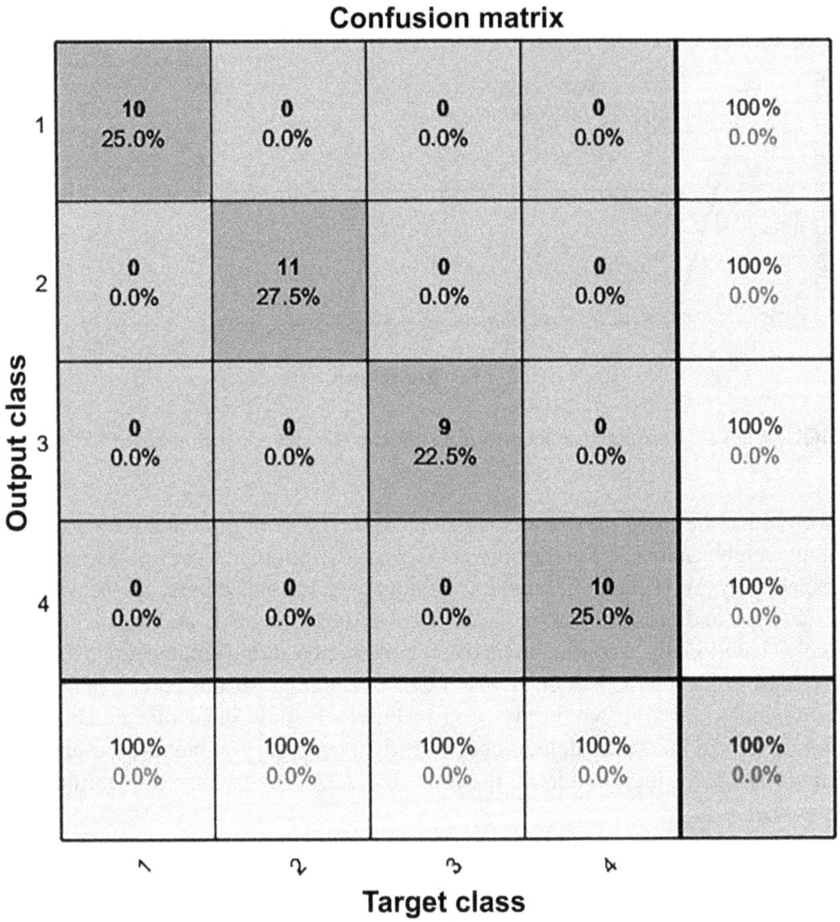

Confusion matrix

FIGURE 5.13   Confusion matrix showing recognition performance of the optimized ELM.

across all health conditions. Additionally, it has been noted that specific defect types can be identified by inputting the data from any health condition into the trained ELM model. The defect identification accuracy for each health condition is presented in Figure 5.14.

To showcase the effectiveness of the algorithm, various optimization methods—including the Ant Lion Optimizer (ALO), Sine-Cosine Algorithm (SCA), Salp Swarm Algorithm (SSA), Grey Wolf Optimization (GWO), and opposition-based PSO—are compared with the proposed opposition-based SMA during the optimization of ELM parameters. The results, evaluated based on accuracy, are displayed in Figure 5.15. The proposed opposition-based SMA achieves the highest accuracy of 100% when optimizing the ELM parameters.

To evaluate the appropriateness of the basis chosen for identifying the dominant mode from the different decomposed modes of SWD, a comparison is made

TABLE 5.3
Comparison of Results of Different Classification Methods

S. No.	Classification Method	Without Feature Ranking		With Feature Ranking	
		Testing Accuracy %	Training Accuracy %	Testing Accuracy %	Training Accuracy %
1.	KNN	76.00	-	77.50	-
2.	SVM	85.00	-	87.50	-
3.	Decision tree	80%	-	83%	-
4.	Random forest	82%	-	86%	-
5.	Proposed method (ELM)	97.65	98.50	100	100

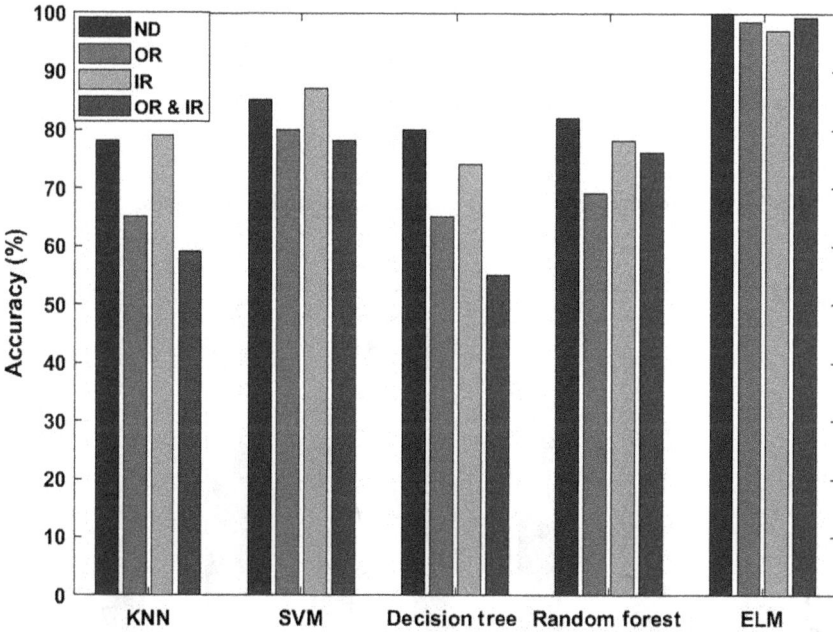

FIGURE 5.14 Defect identification accuracy of the algorithm with different classifiers.

among several entropies, including Shannon Entropy (SE), Sample Entropy (Sp.E), Corrected Conditional Entropy (CCE), Wavelet Energy (WE), and Multiscale PE (MPE) against PE. The results of this comparison are illustrated in Figure 5.16. PE demonstrates superior performance compared to the other measurement indices, thereby establishing it as the basis for selecting the prominent mode.

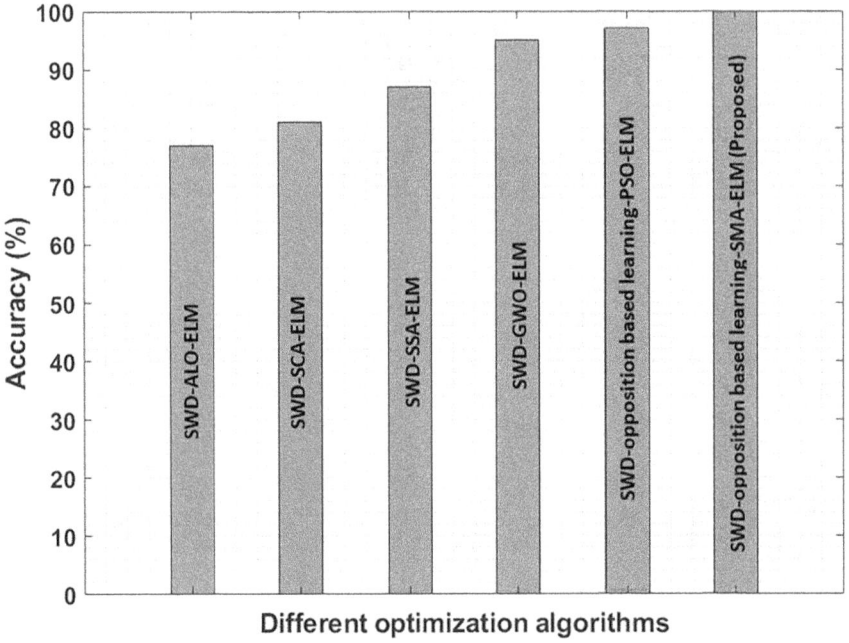

FIGURE 5.15   Accuracy in results with different optimization algorithms.

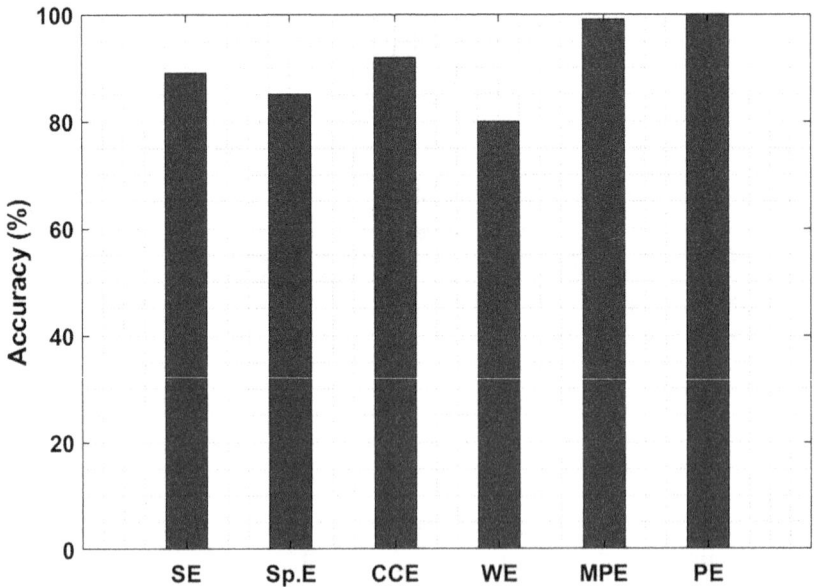

FIGURE 5.16   Comparison of different entropies for selecting prominent mode after decomposition in terms of defect identification accuracy.

5.6 CONCLUSION

The ELM model has been created to identify defects in taper roller bearings. In the signal processing stage, the raw vibration signal acquired from the bearing system is broken down into several modes using SWD. The mode with the lowest PE is selected as the dominant mode. Fault features are identified and prioritized using a filter-based RA. An opposition-based SMA, inspired by nature, is introduced to optimize the parameters of the ELM and to build the classification model. The performance of the developed classifier is evaluated using a fitness function.

The proposed method for identifying bearing defects has been evaluated against other training methods in terms of training accuracy for both ranked and unranked features. It was shown that the overall recognition rate reached 100%, with testing accuracy also recorded at 100%. The experimental outcomes suggest that the proposed approach can effectively identify bearing defects automatically. Furthermore, the fault identification method has been assessed against a variety of signal processing techniques, measurement indices, optimization algorithms, and classifiers to confirm each phase of the process. The proposed fault identification algorithm surpassed all comparisons in terms of accuracy, computational efficiency, sensitivity, specificity, and precision. The results and analysis from this study underscore the benefits of the opposition-based SMA in improving the performance of ELM. This methodology holds considerable promise for use in predictive maintenance within various industrial sectors.

6 The Future of Machine Learning in Fault Diagnosis

6.1 INTRODUCTION

Fault diagnosis, the process of identifying and isolating malfunctions within a system, is crucial across numerous industries, from manufacturing and aerospace to healthcare and energy. Conventional methods for fault diagnosis typically depend on the expertise of professionals, rule-driven systems, and models based on physical principles. However, these approaches can be time-consuming, expensive, and limited in their ability to handle complex systems with high dimensionality and noisy data. The advent of machine learning (ML) has revolutionized fault diagnosis, offering powerful tools to analyze vast datasets, identify subtle patterns, and predict failures with greater accuracy and efficiency. This chapter investigates the prospects of ML in fault diagnosis, analyzing present trends, new technologies on the horizon, and possible obstacles.

6.2 CURRENT LANDSCAPE OF ML IN FAULT DIAGNOSIS

Several ML techniques have proven effective in fault diagnosis applications. These include the following sections.

6.2.1 SUPERVISED LEARNING

This method entails training a model using labeled data, where each instance is linked to a specific fault type. Popular algorithms for this task include support vector machines (SVMs), k-nearest neighbors (KNN), and different varieties of neural networks such as multilayer perceptrons (MLPs), Convolutional neural networks (CNNs), and recurrent neural networks (RNNs). Supervised learning is particularly effective when there is an ample supply of labeled data; however, acquiring this data can be costly and labor intensive.

6.2.2 UNSUPERVISED LEARNING

In situations where labeled data is limited or not accessible, unsupervised learning methods such as clustering (including k-means and density-based spatial clustering of applications with noise (DBSCAN)) and dimensionality reduction

 DOI: 10.1201/9781003614821-6

techniques (such as Principal Component Analysis (PCA) and t-distributed Stochastic Neighbor Embedding (t-SNE)) can be used. These approaches focus on uncovering underlying structures and patterns within the data, which can highlight potential anomalies that suggest faults. Nonetheless, interpreting the outcomes can be difficult, and pinpointing specific fault types may necessitate additional analysis.

6.2.3 SEMI-SUPERVISED LEARNING

This method integrates both labeled and unlabeled data to boost the effectiveness of models trained with a small amount of labeled data. Semi-supervised learning methods utilize the insights present in the unlabeled data to improve the model's ability to generalize.

6.2.4 REINFORCEMENT LEARNING (RL)

Reinforcement learning (RL) is gaining traction in the area of fault diagnosis, particularly within dynamic systems. An RL agent acquires knowledge by engaging with the system, making decisions based on its observations, and receiving rewards or penalties according to how well it identifies and addresses faults. This method holds significant potential for adaptive fault diagnosis in intricate, changing environments.

6.3 EMERGING TRENDS AND TECHNOLOGIES

The field of ML in fault diagnosis is constantly evolving, with several exciting trends and technologies shaping its future.

6.3.1 DEEP LEARNING

Deep learning models, especially deep neural networks (DNNs), have shown remarkable effectiveness in tackling intricate fault diagnosis challenges. CNNs are particularly adept at managing image data from devices like cameras and scanners, whereas RNNs are ideal for examining sequential data from time-series sensors. Recent progress in deep learning, including generative adversarial networks (GANs) and transformers, is also being utilized in fault diagnosis. GANs can produce synthetic data to enhance limited datasets, while transformers excel at managing long-range dependencies in sequential information.

6.3.2 TRANSFER LEARNING

Transfer learning refers to using insights gained from addressing one issue to enhance performance on a similar task. This approach is particularly useful in fault diagnosis when there is insufficient data for a specific system or fault category. By initially training a model on a substantial, relevant dataset and subsequently refining it on the target dataset, the model can attain improved performance with reduced amounts of training data.

6.3.3 EXPLAINABLE AI (XAI)

A key challenge associated with ML models, especially deep learning ones, is their "black box" characteristic. Understanding how these models generate their predictions can be challenging, which makes it difficult to have confidence in their decisions for critical applications like fault diagnosis. XAI techniques are designed to enhance the transparency and interpretability of ML models, facilitating a better understanding of the reasoning behind their predictions. This greater transparency is essential for fostering trust and acceptance of ML-based fault diagnosis systems.

6.3.4 FEDERATED LEARNING

Federated learning enables various parties to jointly develop a common ML model without the need to share their data directly. This approach is especially applicable in fault diagnosis situations where data may be spread across different entities or remote locations due to concerns about privacy or issues related to data ownership.

6.3.5 HYBRID APPROACHES

Integrating ML with conventional techniques, such as physics-based models and expert insights, can result in more dependable and effective fault diagnosis systems. Hybrid methodologies take advantage of the benefits of both ML and traditional approaches, overcoming weaknesses and improving performance. For instance, a physics-based model could deliver preliminary estimates, which are then fine-tuned by a ML model utilizing sensor data.

6.3.6 IoT AND EDGE COMPUTING

The growing implementation of the Internet of Things (IoT) and edge computing is revolutionizing fault diagnosis. IoT devices produce large quantities of data from sensors integrated into diverse systems, offering valuable insights for ML models. Edge computing enables real-time data processing nearer to the source, minimizing latency and bandwidth needs. This supports quicker fault identification and response, which is essential in applications where timing is critical.

6.4 CHALLENGES AND FUTURE DIRECTIONS

Despite the significant progress, several challenges remain in the application of ML to fault diagnosis.

6.4.1 DATA SCARCITY AND QUALITY

Obtaining sufficient high-quality labeled data for training ML models can be challenging and expensive, especially for rare or infrequent fault events. Data augmentation techniques, synthetic data generation, and transfer learning can help mitigate this issue.

6.4.2 DATA HETEROGENEITY AND NOISE

Real-world datasets frequently include noise, incomplete data, and discrepancies, which can impact the effectiveness of ML models. Employing effective data preprocessing and feature engineering methods is essential to tackle these issues.

6.4.3 MODEL INTERPRETABILITY AND EXPLAINABILITY

The lack of transparency in some ML models makes it difficult to understand their predictions, hindering trust and acceptance. XAI methods are essential for addressing this issue and improving confidence in ML-based fault diagnosis systems.

6.4.4 GENERALIZATION AND ROBUSTNESS

ML models trained on one dataset might not generalize well to other datasets or operating conditions. Developing robust and generalizable models is crucial for reliable fault diagnosis across various scenarios.

6.4.5 REAL-TIME PERFORMANCE AND SCALABILITY

For some applications, real-time fault detection and diagnosis are essential. Developing ML models that can operate efficiently and scale to handle large datasets in real time is a critical challenge.

6.4.6 SECURITY AND SAFETY

Implementing ML models in essential systems demands thorough attention to safety and security factors. It is crucial to guarantee the strength and dependability of these models to avoid unexpected outcomes.

6.5 SPECIFIC APPLICATION AREAS AND FUTURE OUTLOOK

The future of ML in fault diagnosis holds immense potential across various domains.

6.5.1 MANUFACTURING

ML has the potential to transform predictive maintenance, minimizing downtime and enhancing efficiency in manufacturing operations. By examining sensor data from equipment, ML models can forecast possible failures and arrange maintenance in advance.

6.5.2 AEROSPACE

In aerospace contexts, diagnosing faults is essential for maintaining safety and reliability. ML can enhance the precision and speed of fault identification in aircraft and spacecraft systems, which may help avert disastrous failures.

6.5.3 HEALTHCARE

ML has the potential to improve medical diagnosis by examining medical images, patient information, and other pertinent data. This advancement can result in quicker and more precise disease identification, ultimately benefiting patient outcomes.

6.5.4 ENERGY

ML can enhance the effectiveness and dependability of power grids and various energy systems by identifying and diagnosing faults in real time. This capability can aid in preventing power outages and optimizing energy management.

6.5.5 AUTOMOTIVE

Self-driving cars depend significantly on effective fault diagnosis mechanisms to guarantee safety and dependability. ML can enhance the precision and rapidity of fault detection in autonomous driving technologies, reducing the likelihood of accidents and boosting overall efficiency.

6.6 CONCLUSION

ML is revolutionizing fault diagnosis in a variety of industries by providing robust tools that enhance accuracy, efficiency, and reliability. New trends such as deep learning, transfer learning, XAI, federated learning, and hybrid methods are further augmenting the capabilities of ML-based fault diagnosis systems. Nevertheless, issues related to data limitation, diversity, model transparency, and real-time performance must be tackled to fully harness the potential of ML in this area. Current research and development initiatives are aimed at addressing these challenges and broadening the use of ML in increasingly complex and critical systems. The future of fault diagnosis is closely linked to the advancements in ML, resulting in safer, more efficient, and more dependable systems across various sectors. Ongoing partnerships between researchers, engineers, and industry professionals will be crucial for advancing this swiftly developing field.

References

[1] Collacott RA. Mechanical fault diagnosis. *Shipcare Int (Formerly Ship Repair and Maihtenance)* 1976;8.

[2] Kryter RC. *Condition Monitoring of machinery using MCSA*; 1989.

[3] Tiwari M, Gupta K, Prakash O. Dynamic response of an unbalanced rotor supported on ball bearings. *Journal of Sound and Vibration* 2000;238:757–79. https://doi.org/10.1006/jsvi.1999.3108

[4] Bajda M, Blazej R, Jurdziak L, Kirjanow A. Condition monitoring of textile belts in the light of research results of their resistance to punctures investigations. *16th International Multidisciplinary Scientific GeoConference SGEM2016, Science and Technologies in Geology, Exploration and Mining*, vol. 2, 2016. https://doi.org/10.5593/sgem2016/b12/s03.022

[5] Mba D, Rao RB. Development of acoustic emission technology for condition monitoring and diagnosis of rotating machines: Bearings, pumps, gearboxes, engines, and rotating structures. *Shock and Vibration Digest* 2006;38:3–16. https://doi.org/10.1177/0583102405059054

[6] Randall RB. *Vibration-Based Condition Monitoring: Industrial, Aerospace and Automotive Applications*. Wiley; 2011.

[7] Swansson NS, Favaloro SC. Applications od vibration analysis to the condition monitoring of rolling element bearings. *Lubrication, Friction and Wear in Engineering 1980, Melbourne: Preprints of Papers*. Institution of Engineers, Australia; 1980.

[8] Batchelor BG, Hill DA, Hodgson DC. *Automated Visual Inspection*. Kempston, Bedford: IFS; 1985.

[9] Tiwari M, Gupta K, Prakash O. Effect of radial internal clearance of a ball bearing on the dynamics of a balanced horizontal rotor. *Journal of Sound and Vibration* 2000;238:723–56. https://doi.org/10.1006/jsvi.1999.3109

[10] Jiang J. *Acoustic Condition Monitoring in Industrial Environments: A Book Renewing Your Understanding of Tradiational Condition Monitoring*. Germany LAP Lambert: Saarbrucken; 2011.

[11] Singh A, Parey A. Gearbox fault diagnosis under non-stationary conditions with independent angular re-sampling technique applied to vibration and sound emission signals. *Applied Acoustics* 2019;144:11–22. https://doi.org/10.1016/j.apacoust.2017.04.015

[12] Sachse W, Roget J, Yamaguchi K. *Acoustic Emission: Current Practice and Future Directions*; 1991. https://doi.org/10.1520/stp1077-eb

[13] Yan SF, Ma B, Zheng CS, Chen M. Weighted evidential fusion method for fault diagnosis of mechanical transmission based on oil analysis data. *International Journal of Automatic Technology* 2019;20:989–96. https://doi.org/10.1007/s12239-019-0093-9

[14] Gubran HBH, Singh SP, Gupta K. Stresses in composite shafts subjected to unbalance excitation and transmitted torque. *International Journal of Rotating Machinery* 2000;6:235–44. https://doi.org/10.1155/S1023621X00000221

[15] Prabhakar S, Mohanty AR, Sekhar AS. Crack versus coupling misalignment in a transient rotor system. *Journal of Sound and Vibration* 2002;256:773–86.

[16] de Moura EP, Junior FE, Damasceno FF, Figueiredo LC, de Andrade CF, de Almeida MS, Rocha PA. Classification of imbalance levels in a scaled wind turbine through detrended fluctuation analysis of vibration signals. *Renewable Energy* 2016;96:993–1002. https://doi.org/10.1016/j.renene.2016.05.005

[17] Ferrando Chacon JL, Artigao Andicoberry E, Kappatos V, Asfis G, Gan TH, Balachandran W. Shaft angular misalignment detection using acoustic emission. *Applied Acoustics* 2014;85:12–22. https://doi.org/10.1016/j.apacoust.2014.03.018

[18] Lal M. Modeling and estimation of speed dependent bearing and coupling misalignment faults in a turbine generator system. *Mechanical Systems and Signal Processing* 2021;151:107365. https://doi.org/10.1016/j.ymssp.2020.107365

[19] The Construction Building Ideas. Franicis Turbine-its Components, Working and Application 2022. https://theconstructor.org/practical-guide/francis-turbines-components-application/2900/ (accessed June 21, 2022).

[20] Padhy MK, Saini RP. Study of silt erosion mechanism in Pelton turbine buckets. *Energy* 2012;39:286–93. https://doi.org/10.1016/j.energy.2012.01.015

[21] Rai AK, Kumar A, Staubli T. Forces acting on particles in a Pelton bucket and similarity considerations for erosion. *IOP Conference Series: Earth and Environmental Science* 2016;49. https://doi.org/10.1088/1755-1315/49/12/122002

[22] Rai AK, Kumar A, Staubli T. Hydro-abrasive erosion in Pelton buckets: Classification and field study. *Wear* 2017;392–393:8–20. https://doi.org/10.1016/j.wear.2017.08.016

[23] Rossetti A, Pavesi G, Ardizzon G, Santolin A. Numerical analyses of cavitating flow in a pelton turbine. *Journal of Fluids Engineering* 2014;136:081304. https://doi.org/10.1115/1.4027139

[24] Rai AK, Kumar A, Staubli T. Forces acting on particles in a Pelton bucket and similarity considerations for erosion. *IOP Conference Series: Earth and Environmental Science* 2016;49. https://doi.org/10.1088/1755-1315/49/12/122002

[25] Harris TA, Kotzalas MN. *Advanced Concepts of Bearing Technology.* 5th ed. CRC; 2006.

[26] Orhan S, Aktürk N, Çelik V. Vibration monitoring for defect diagnosis of rolling element bearings as a predictive maintenance tool: Comprehensive case studies. *NDT and E International* 2006;39:293–8. https://doi.org/10.1016/j.ndteint.2005.08.008

[27] Parey A, El Badaoui M, Guillet F, Tandon N. Dynamic modelling of spur gear pair and application of empirical mode decomposition-based statistical analysis for early detection of localized tooth defect. *Journal of Sound and Vibration* 2006;294:547–61. https://doi.org/10.1016/j.jsv.2005.11.021

[28] Wu S, Zuo MJ, Parey A. Simulation of spur gear dynamics and estimation of fault growth. *Journal of Sound and Vibration* 2008;317:608–24. https://doi.org/10.1016/j.jsv.2008.03.038

[29] Singh J, Darpe AK, Singh SP. Bearing damage assessment using Jensen-Rényi Divergence based on EEMD. *Mechanical Systems and Signal Processing* 2017;87:307–39. https://doi.org/10.1016/j.ymssp.2016.10.028

[30] Vernekar K, Kumar H, Gangadharan KV. Engine gearbox fault diagnosis using empirical mode decomposition method and Naïve Bayes algorithm. *Sadhana - Academy Proceedings in Engineering Sciences* 2017;42:1143–53. https://doi.org/10.1007/s12046-017-0678-9

[31] Kankar PK, Sharma SC, Harsha SP. Vibration based performance prediction of ball bearings caused by localized defects. *Nonlinear Dynamics* 2012;69:847–75. https://doi.org/10.1007/s11071-011-0309-7

[32] Roth JT, Djurdjanovic D, Yang X, Mears L, Kurfess T. Quality and inspection of machining operations: Tool condition monitoring. *Journal of Manufacturing Science and Engineering, Transactions of the ASME* 2010;132:0410151–04101516. https:// doi.org/10.1115/1.4002022

[33] Egusquiza M, Egusquiza E, Valentin D, Valero C, Presas A. Failure investigation of a Pelton turbine runner. *Engineering Failure Analysis* 2017;81:234–44. https://doi. org/10.1016/j.engfailanal.2017.06.048

[34] Egusquiza M, Egusquiza E, Valero C, Presas A, Valentín D, Bossio M. Advanced condition monitoring of Pelton turbines. *Measurement* 2018;119:46–55. https://doi. org/10.1016/j.measurement.2018.01.030

[35] Carp-Ciocardia D-C, Safta C, Dragomirescu A, Magheti I, Schiaua M. Experimental analysis of vibrations at a pelton turbine. *Proceedings of the Annual Symposium of the Institute of Solid Mechanics and Session of Commission of Acoustics* 2018;59:7–10.

[36] Kumar A, Kumar R. Time-frequency analysis and support vector machine in automatic detection of defect from vibration signal of centrifugal pump. *Measurement* 2017;108:119–33. https://doi.org/10.1016/j.measurement.2017.04.041

[37] Zhao W, Egusquiza M, Valero C, Valentín D, Presas A, Egusquiza E. On the use of artificial neural networks for condition monitoring of pump-turbines with extended operation. *Measurement* 2020;163:107952. https://doi.org/10.1016/j.measurement.2020.107952

[38] Feng Z, Liang M, Zhang Y, Hou S. Fault diagnosis for wind turbine planetary gearboxes via demodulation analysis based on ensemble empirical mode decomposition and energy separation. *Renewable Energy* 2012;47:112–26. https://doi.org/10.1016/j. renene.2012.04.019

[39] Mohan TR, Roselyn JP, Uthra RA, Devaraj D, Umachandran K, Intelligent machine learning based total productive maintenance approach for achieving zero downtime in industrial machinery. *Computers and Industrial Engineering* 2021;157:107267. https://doi.org/10.1016/j.cie.2021.107267

[40] Feng Z, Ma H, Zuo MJ. Vibration signal models for fault diagnosis of planet bearings. *Journal of Sound and Vibration* 2016;370:372–93. https://doi.org/10.1016/j. jsv.2016.01.041

[41] Rodopoulos KI, Antoniadis IA. Instantaneous fault frequencies estimation in roller bearings via wavelet structures. *Journal of Sound and Vibration* 2016;383:446–63. https://doi.org/10.1016/j.jsv.2016.07.027

[42] Wang D, Miao Q, Fan X, Huang H. Rolling element bearing fault detection using an improved combination of Hilbert and Wavelet transforms. *Journal of Mechanical Science and Technology* 2009;23:3292–301. https://doi.org/10.1007/ s12206-009-0807-4

[43] Li Z, Feng Z, Chu F. A load identification method based on wavelet multi-resolution analysis. *Journal of Sound and Vibration* 2014;333:381–91. https://doi.org/10.1016/j. jsv.2013.09.026

[44] Cui L, Gong X, Zhang J, Wang H. Double-dictionary matching pursuit for fault extent evaluation of rolling bearing based on the Lempel-Ziv complexity. *Journal of Sound and Vibration* 2016;385:372–88. https://doi.org/10.1016/j.jsv.2016.09.008

[45] Huang G, Xiao Y, Yin Z. Denoising method for underwater acoustic signals based on sparse decomposition. *Journal of Physics: Conference Series* 2020;1550. https://doi. org/10.1088/1742-6596/1550/3/032139

[46] Liu Z, He Z, Guo W, Tang Z. A hybrid fault diagnosis method based on second generation wavelet de-noising and local mean decomposition for rotating machinery. *ISA Transactions* 2016;61:211–20. https://doi.org/10.1016/j.isatra.2015.12.009

[47] Zhang X, Wang L, Miao Q. Fault diagnosis techniques for planetary gearboxes under variable conditions: A review. *Proceedings of 2016 Prognostics and System Health Management Conference, PHM-Chengdu 2016* 2017:1–11. https://doi.org/10.1109/PHM.2016.7819889

[48] Guo L, Li N, Jia F, Lei Y, Lin J. A recurrent neural network based health indicator for remaining useful life prediction of bearings. *Neurocomputing* 2017;240:98–109. https://doi.org/10.1016/j.neucom.2017.02.045

[49] Huang NE, Shen Z, Long SR, Wu MC, Snin HH, Zheng Q, et al. The empirical mode decomposition and the Hubert spectrum for nonlinear and non-stationary time series analysis. *Proceedings of the Royal Society A: Mathematical, Physical and Engineering Sciences* 1998;454:903–95. https://doi.org/10.1098/rspa.1998.0193

[50] Chauhan S, Singh M, Kumar Aggarwal A. An effective health indicator for bearing using corrected conditional entropy through diversity-driven multi-parent evolutionary algorithm. *Structural Health Monitoring* 2020;20(5), 2525–39.

[51] Li H, Li Z, Mo W. A time varying filter approach for empirical mode decomposition. *Signal Processing* 2017;138:146–58. https://doi.org/10.1016/j.sigpro.2017.03.019

[52] Wang L, Shao Y. Fault feature extraction of rotating machinery using a reweighted complete ensemble empirical mode decomposition with adaptive noise and demodulation analysis. *Mechanical Systems and Signal Processing* 2020;138:106545. https://doi.org/10.1016/j.ymssp.2019.106545

[53] Rehman N, Mandic DP. Multivariate empirical mode decomposition. *Proceedings of the Royal Society A: Mathematical, Physical and Engineering Sciences* 2010;466:1291–302. https://doi.org/10.1098/rspa.2009.0502

[54] Zhang X, Liu Z, Miao Q, Wang L. An optimized time varying filtering based empirical mode decomposition method with grey wolf optimizer for machinery fault diagnosis. *Journal of Sound and Vibration* 2018;418:55–78. https://doi.org/10.1016/j.jsv.2017.12.028

[55] Seventekidis P, Giagopoulos D, Arailopoulos A, Markogiannaki O. Structural Health Monitoring using deep learning with optimal finite element model generated data. *Mechanical Systems and Signal Processing* 2020;145:106972. https://doi.org/10.1016/j.ymssp.2020.106972

[56] Kumar A, Gandhi CP, Zhou Y, Kumar R, Xiang J. Improved deep convolution neural network (CNN) for the identification of defects in the centrifugal pump using acoustic images. *Applied Acoustics* 2020;167:107399. https://doi.org/10.1016/j.apacoust.2020.107399

[57] Margolin AA, Nemenman I, Basso K, Wiggins C, Stolovitzky G, Favera RD, et al. ARACNE: An algorithm for the reconstruction of gene regulatory networks in a mammalian cellular context. *BMC Bioinformatics* 2006;15:1–15. https://doi.org/10.1186/1471-2105-7-S1-S7

[58] Kumar A, Zhou Y, Xiang J. Optimization of VMD using kernel-based mutual information for the extraction of weak features to detect bearing defects. *Measurement* 2021;168:108402. https://doi.org/10.1016/j.measurement.2020.108402

[59] Mirjalili S, Mirjalili SM, Lewis A. Grey wolf optimizer. *Advances in Engineering Software* 2014;69:46–61. https://doi.org/10.1016/j.advengsoft.2013.12.007

[60] Bäck T, Schwefel H-P. An overview of evolutionary algorithms for parameter optimization. *Evolutionary Computation* 1993;1:1–23. https://doi.org/10.1162/evco.1993.1.1.1

[61] Luo J, Chen H, Zhang Q, Xu Y, Huang H, Zhao X. An improved grasshopper optimization algorithm with application to financial stress prediction. *Applied Mathematical Modelling* 2018;64:654–68. https://doi.org/10.1016/j.apm.2018.07.044

[62] Salgotra R, Singh U. Application of mutation operators to flower pollination algorithm. *Expert Systems with Applications* 2017;79:112–29. https://doi.org/10.1016/j.eswa.2017.02.035

[63] Zhang M, Valentin D, Valero C, Egusquiza M, Zhao W. Numerical study on the dynamic behavior of a francis turbine runner model with a crack. *Energies* 2018;11. https://doi.org/10.3390/en11071630

[64] Thapa BS, Dahlhaug OG, Thapa B. Sediment erosion induced leakage flow from guide vane clearance gap in a low specific speed Francis turbine. *Renewable Energy* 2017;107:253–61. https://doi.org/10.1016/j.renene.2017.01.045

[65] Zhang H, Zhang L. Numerical simulation of cavitating turbulent flow in a high head Francis turbine at part load operation with OpenFOAM. *Procedia Engineering* 2012;31:156–65. https://doi.org/10.1016/j.proeng.2012.01.1006

[66] Valentín D, Presas A, Valero C, Egusquiza M, Egusquiza E, Gomes J, et al. Transposition of the mechanical behavior from model to prototype of Francis turbines. *Renewable Energy* 2020;152:1011–23. https://doi.org/10.1016/j.renene.2020.01.115

[67] Kumar A, Gandhi CP, Zhou Y, Kumar R, Xiang J. Improved deep convolution neural network (CNN) for the identification of defects in the centrifugal pump using acoustic images. *Applied Acoustics* 2020;167:107399. https://doi.org/10.1016/j.apacoust.2020.107399

[68] Kumar A, Gandhi CP, Zhou Y, Kumar R, Xiang J. Variational mode decomposition based symmetric single valued neutrosophic cross entropy measure for the identification of bearing defects in a centrifugal pump. *Applied Acoustics* 2020;165:107294. https://doi.org/10.1016/j.apacoust.2020.107294

[69] Sharma RK, Pandey RK. Experimental studies of pressure distributions in finite slider bearing with single continuous surface profiles on the pads. *Tribology International* 2009;42:1040–5. https://doi.org/10.1016/j.triboint.2009.02.010

[70] Liu J, Zhang Q, Yin Z, Sun C, Chen X, Song Z. An enhanced adaptive notch filtering method for online multi-frequency estimation from contaminated signals of a mechanical control system. *Measurement Science and Technology* 2021;32:105102.

[71] Kumar TP, Saimurugan M, Hari Haran RB, Siddharth S, Ramachandran KI. A multi-sensor information fusion for fault diagnosis of a gearbox utilizing discrete wavelet features. *Measurement Science and Technology* 2019;30:085105. https://doi.org/10.1080/14484846.2018.1432089

[72] Liu X, Hu Z, He Q, Zhang S, Zhu J. Doppler distortion correction based on microphone array and matching pursuit algorithm for a wayside train bearing monitoring system. *Measurement Science and Technology* 2017;28:105006. https://doi.org/10.1088/1361-6501/aa67c8

[73] Yuan Z, Zhang L, Duan L. A novel fusion diagnosis method for rotor system fault based on deep learning and multi-sourced heterogeneous monitoring data a novel fusion diagnosis method for rotor system faults based on deep lear. *Measurement Science and Technology* 2018;29:115005.

[74] Wiggins RA. Maximum entropy deconvolution. *Geoexploration* 1978;16:21–35. https://doi.org/10.1109/icassp.1986.1169086

[75] Sawalhi N, Randall RB, Endo H. The enhancement of fault detection and diagnosis in rolling element bearings using minimum entropy deconvolution combined with spectral kurtosis. *Mechanical Systems and Signal Processing* 2007;21:2616–33.

[76] Wang S, Xiang J, Tang H, Liu X, Zhong Y. Minimum entropy deconvolution based on simulation-determined band pass filter to detect faults in axial piston pump bearings. *ISA Transactions* 2019;88:186–98. https://doi.org/10.1016/j.isatra.2018.11.040

[77] Endo H, Randall RB. Enhancement of autoregressive model based gear tooth fault detection technique by the use of minimum entropy deconvolution filter. *Mechanical Systems and Signal Processing* 2007;21:906–19. https://doi.org/10.1016/j. ymssp.2006.02.005

[78] Gao Y, Karimi M, Kudreyko AA, Song W. Spare optimistic based on improved ADMM and the minimum entropy de-convolution for the early weak fault diagnosis of bearings in marine systems. *ISA Transactions* 2018;78:98–104. https://doi. org/10.1016/j.isatra.2017.12.021

[79] Ovacıklı AK, Member S, Pääjärvi P, Leblanc JP, Member S, Carlson JE. Recovering periodic impulsive signals through skewness maximization. *IEEE Transactions on Signal Processing* 2016;64:1586–96.

[80] Obuchowski J, Zimroz R, Wyłomańska A. Blind equalization using combined skewness-kurtosis criterion for gearbox vibration enhancement. *Measurement: Journal of the International Measurement Confederation* 2016;88:34–44. https:// doi.org/10.1016/j.measurement.2016.03.034

[81] McDonald GL, Zhao Q, Zuo MJ. Maximum correlated Kurtosis deconvolution and application on gear tooth chip fault detection. *Mechanical Systems and Signal Processing* 2012;33:237–55. https://doi.org/10.1016/j.ymssp.2012.06.010

[82] McDonald GL, Zhao Q. Multipoint optimal minimum entropy deconvolution and convolution fix: Application to vibration fault detection. *Mechanical Systems and Signal Processing* 2017;82:461–77. https://doi.org/10.1016/j.ymssp.2016.05.036

[83] Cheng Y, Chen B, Mei G, Wang Z, Zhang W. A novel blind deconvolution method and its application to fault identification. *Journal of Sound and Vibration* 2019;460:114900. https://doi.org/10.1016/j.jsv.2019.114900

[84] Zheng K, Luo J, Zhang Y, Li T, Wen J, Xiao H. Incipient fault detection of rolling bearing using maximum autocorrelation impulse harmonic to noise deconvolution and parameter optimized fast EEMD. *ISA Transactions* 2019;89:256–71. https://doi. org/10.1016/j.isatra.2018.12.020

[85] Cheng Y, Zhou N, Zhang W, Wang Z. Application of an improved minimum entropy deconvolution method for railway rolling element bearing fault diagnosis. *Journal of Sound and Vibration* 2018;425:53–69. https://doi.org/10.1016/j.jsv.2018.01.023

[86] Jiang X, Cheng X, Shi J, Huang W, Shen C, Zhu Z. A new l0-norm embedded MED method for roller element bearing fault diagnosis at early stage of damage. *Measurement* 2018;127:414–24. https://doi.org/10.1016/j.measurement.2018.06.016

[87] Barszcz T, Sawalhi N. Fault detection enhancement in rolling element bearings using the minimum entropy deconvolution. *Archives of Acoustics* 2012;37:131–41. https:// doi.org/10.2478/v10168-012-0019-2

[88] Miao Y, Zhao M, Lin J, Lei Y. Application of an improved maximum correlated kurtosis deconvolution method for fault diagnosis of rolling element bearings. *Mechanical Systems and Signal Processing* 2017;92:173–95. https://doi. org/10.1016/j.ymssp.2017.01.033

[89] Li J, Li M, Zhang J. Rolling bearing fault diagnosis based on time-delayed feedback monostable stochastic resonance and adaptive minimum entropy deconvolution. *Journal of Sound and Vibration* 2017;401:139–51. https://doi.org/10.1016/j. jsv.2017.04.036

[90] Zhang L, Hu N. Fault diagnosis of sun gear based on continuous vibration separation and minimum entropy deconvolution. *Measurement* 2019;141:332–44. https://doi. org/10.1016/j.measurement.2019.04.049

[91] He D, Wang X, Li S, Lin J, Zhao M. Identification of multiple faults in rotating machinery based on minimum entropy deconvolution combined with spectral kurtosis. *Mechanical Systems and Signal Processing* 2016;81:235–49. https://doi.org/10.1016/j.ymssp.2016.03.016

[92] Li G, Zhao Q. Minimum entropy deconvolution optimized sinusoidal synthesis and its application to vibration based fault detection. *Journal of Sound and Vibration* 2017;390:218–31. https://doi.org/10.1016/j.jsv.2016.11.033

[93] Mirjalili S, Gandomi AH, Mirjalili SZ, Saremi S, Faris H, Mirjalili SM. Salp swarm algorithm: A bio-inspired optimizer for engineering design problems. *Advances in Engineering Software* 2017;114:163–91. https://doi.org/10.1016/j.advengsoft.2017.07.002

[94] Al-Shourbaji I, Helian N, Sun Y, Alshathri S, Elaziz MA. Boosting ant colony optimization with reptile search algorithm for churn prediction. *Mathematics* 2022;10:1–21. https://doi.org/10.3390/math10071031

[95] Nadimi-Shahraki MH, Taghian S, Mirjalili S. An improved grey wolf optimizer for solving engineering problems. *Expert Systems with Applications* 2021;166:113917. https://doi.org/10.1016/j.eswa.2020.113917

[96] Abualigah L, Yousri D, Abd Elaziz M, Ewees AA, Al-Qaness MAA, Gandomi AH. Aquila optimizer: A novel meta-heuristic optimization algorithm. *Computers and Industrial Engineering* 2021;157:107250. https://doi.org/10.1016/j.cie.2021.107250

[97] McDonald GL, Zhao Q. Multipoint optimal minimum entropy deconvolution and convolution fix: Application to vibration fault detection. *Mechanical Systems and Signal Processing* 2017;82:461–77. https://doi.org/10.1016/j.ymssp.2016.05.036

[98] Tiwari R, Bordoloi DJ, Dewangan A. Blockage and cavitation detection in centrifugal pumps from dynamic pressure signal using deep learning algorithm. *Measurement* 2020;173:108676. https://doi.org/10.1016/j.measurement.2020.108676

[99] Kumar A, Kumar R. Oscillatory behavior-based wavelet decomposition for the monitoring of bearing condition in centrifugal pumps. *Proceedings of the Institution of Mechanical Engineers, Part J: Journal of Engineering Tribology* 2018;232:757–72. https://doi.org/10.1177/1350650117727976

[100] Zhao X, Zuo MJ, Patel TH. Generating an indicator for pump impeller damage using half and full spectra, fuzzy preference-based rough sets and PCA. *Measurement Science and Technology* 2012;23. https://doi.org/10.1088/0957-0233/23/4/045607

[101] Kumar A, Kumar R. Least square fitting for adaptive wavelet generation and automatic prediction of defect size in the bearing using levenberg–marquardt backpropagation. *Journal of Nondestructive Evaluation* 2017;36. https://doi.org/10.1007/s10921-016-0385-1

[102] Kumar A, Kumar R. Adaptive artificial intelligence for automatic identification of defect in the angular contact bearing. *Neural Computing and Applications* 2018;29:277–87. https://doi.org/10.1007/s00521-017-3123-4

[103] Kumar A, Kumar R. Development of LDA based indicator for the detection of unbalance and misalignment at different shaft speeds. *Experimental Techniques* 2019. https://doi.org/10.1007/s40799-019-00349-5

[104] Rapur JS, Tiwari R. Experimental time-domain vibration- based fault diagnosis of centrifugal pumps using support vector machine. *ASCE-ASME Journal of Risk and Uncertainty in Engineering Systems, Part B: Mechanical Engineering* 2017;3. https://doi.org/10.1115/1.4035440

[105] Zhang M, Jiang Z, Feng K. Research on variational mode decomposition in rolling bearings fault diagnosis of the multistage centrifugal pump. *Mechanical Systems and Signal Processing* 2017;93:460–93. https://doi.org/10.1016/j.ymssp.2017.02.013

[106] Muralidharan V, Sugumaran V, Indira V. Fault diagnosis of monoblock centrifugal pump using SVM. *Engineering Science and Technology, an International Journal* 2014;17:152–7. https://doi.org/10.1016/j.jestch.2014.04.005

[107] Azizi R, Attaran B, Hajnayeb A, Ghanbarzadeh A, Changizian M. Improving accuracy of cavitation severity detection in centrifugal pumps using a hybrid feature selection technique. *Measurement* 2017;108:9–17. https://doi.org/10.1016/j.measurement.2017.05.020

[108] Kumar A, Kumar R. Time-frequency analysis and support vector machine in automatic detection of defect from vibration signal of centrifugal pump. *Measurement* 2017;108:119–33. https://doi.org/10.1016/j.measurement.2017.04.041

[109] Kumar A, Gandhi CP, Zhou Y, Tang H, Xiang J. Fault diagnosis of rolling element bearing based on symmetric cross entropy of neutrosophic sets. *Measurement* 2020;152:107318. https://doi.org/10.1016/j.measurement.2019.107318

[110] Dragomiretskiy K, Zosso D. Variational mode decomposition. *IEEE Transactions on Signal Processing* 2014;62:531–44. https://doi.org/10.1109/TSP.2013.2288675

[111] Chauhan S, Singh M, Kumar Aggarwal A. Bearing defect identification via evolutionary algorithm with adaptive wavelet mutation strategy. *Measurement* 2021;179:109445. https://doi.org/10.1016/j.measurement.2021.109445

[112] Chauhan S, Singh M, Aggarwal AK. Diversity driven multi-parent evolutionary algorithm with adaptive non-uniform mutation. *Journal of Experimental and Theoretical Artificial Intelligence* 2020:1–32. https://doi.org/10.1080/0952813X.2020.1785020

[113] Kumar A, Gandhi CP, Tang H, Vashishtha G, Kumar R, Zhou Y, et al. Adaptive sensitive frequency band selection for VMD to identify defective components of an axial piston pump. *Chinese Journal of Aeronautics* 2021. https://doi.org/10.1016/j.cja.2020.12.037

[114] Huang GB, Zhu QY, Siew CK. Extreme learning machine: Theory and applications. *Neurocomputing* 2006;70:489–501. https://doi.org/10.1016/j.neucom.2005.12.126

[115] Li J, Chen X, He Z. Adaptive stochastic resonance method for impact signal detection based on sliding window. *Mechanical Systems and Signal Processing* 2013;36:240–55. https://doi.org/10.1016/j.ymssp.2012.12.004

[116] Borghesani P, Pennacchi P, Chatterton S. The relationship between kurtosis- and envelope-based indexes for the diagnostic of rolling element bearings. *Mechanical Systems and Signal Processing* 2014;43:25–43. https://doi.org/10.1016/j.ymssp.2013.10.007

[117] Zhang X, Miao Q, Zhang H, Wang L. A parameter-adaptive VMD method based on grasshopper optimization algorithm to analyze vibration signals from rotating machinery. *Mechanical Systems and Signal Processing* 2018;108:58–72. https://doi.org/10.1016/j.ymssp.2017.11.029

[118] Mirjalili S, Gandomi AH, Mirjalili SZ, Saremi S, Faris H, Mirjalili SM. Salp swarm algorithm: A bio-inspired optimizer for engineering design problems. *Advances in Engineering Software* 2017;114:163–91. https://doi.org/10.1016/j.advengsoft.2017.07.002

[119] Biesiada J, Duch W. Feature selection for high-dimensional data – A pearson redundancy based filter. *Advances in Soft Computing* 2007;45:242–9. https://doi.org/10.1007/978-3-540-75175-5_30

[120] Igel C, Toussaint M. A no-free-lunch theorem for non-uniform distributions of target functions. *Journal of Mathematical Modelling and Algorithms* 2004;3:313–22. https://doi.org/10.1023/B:JMMA.0000049381.24625.f7

[121] Ngiam J, Koh PW, Chen Z, Bhaskar S, Ng AY. Sparse filtering. *Advances in Neural Information Processing Systems 24: 25th Annual Conference on Neural Information Processing Systems 2011, NIPS 2011* 2011:1–9.

[122] Lei Y, Jia F, Lin J, Xing S, Ding SX. An intelligent fault diagnosis method using unsupervised feature learning towards mechanical big data. *IEEE Transactions on Industrial Electronics* 2016;63:3137–47. https://doi.org/10.1109/TIE.2016.2519325

[123] Tang B, Song T, Li F, Deng L. Fault diagnosis for a wind turbine transmission system based on manifold learning and Shannon wavelet support vector machine. *Renewable Energy* 2014;62:1–9. https://doi.org/10.1016/j.renene.2013.06.025

[124] Han B, Zhang X, Wang J, An Z, Jia S, Zhang G. Hybrid distance-guided adversarial network for intelligent fault diagnosis under different working conditions. *Measurement* 2021;176:109197. https://doi.org/10.1016/j.measurement.2021.109197

[125] Hochreiter S, Schmidhuber J. Long short-term memory. *Neural Computation* 1997;9:1735–80.

[126] Zhu A, Zhao Q, Yang T, Zhou L, Zeng B. Condition monitoring of wind turbine based on deep learning networks and kernel principal component analysis. *Computers and Electrical Engineering* 2023;105:108538. https://doi.org/10.1016/J.COMPELECENG.2022.108538

[127] Schmidt S, Gryllias KC. An anomalous frequency band identification method utilising available healthy historical data for gearbox fault detection. *Measurement* 2023;222:113515. https://doi.org/10.1016/j.measurement.2023.113515

[128] Singh G, Naikan VNA. Partial broken rotor bar fault diagnosis using signal injected and generated Hilbert method. *Computers and Electrical Engineering* 2023;111:108935. https://doi.org/10.1016/J.COMPELECENG.2023.108935

[129] Vashishtha G, Chauhan S, Kumar A, Kumar R. An ameliorated African vulture optimization algorithm to diagnose the rolling bearing defects. *Measurement Science and Technology* 2022;33.

[130] Vashishtha G, Chauhan S, Yadav N, Kumar A, Kumar R. A two-level adaptive chirp mode decomposition and tangent entropy in estimation of single-valued neutrosophic cross-entropy for detecting impeller defects in centrifugal pump. *Applied Acoustics* 2022;197:108905.

[131] Vashishtha G, Chauhan S, Kumar S, Kumar R, Zimroz R, Kumar A. Intelligent fault diagnosis of worm gearbox based on adaptive CNN using amended gorilla troop optimization with quantum gate mutation strategy. *Knowledge-Based Systems* 2023;280:110984. https://doi.org/10.1016/j.knosys.2023.110984

[132] Li Y, Xu M, Haiyang Z, Wei Y, Huang W. A new rotating machinery fault diagnosis method based on improved local mean decomposition. *Digital Signal Processing: A Review Journal* 2015;46:201–14. https://doi.org/10.1016/j.dsp.2015.07.001

[133] Han D, Zhao N, Shi P. Gear fault feature extraction and diagnosis method under different load excitation based on EMD, PSO-SVM and fractal box dimension. *Journal of Mechanical Science and Technology* 2019;33:487–94. https://doi.org/10.1007/s12206-019-0101-z

[134] Kumar A, Kumar R. Role of signal processing, modeling and decision making in the diagnosis of rolling element bearing defect: A review. *Journal of Nondestructive Evaluation* 2019;123. https://doi.org/10.1007/s10921-018-0543-8

[135] Buzzoni M, Mucchi E, D'Elia G, Dalpiaz G. Diagnosis of localized faults in multi-stage gearboxes: A vibrational approach by means of automatic EMD-based algorithm. *Shock and Vibration* 2017;2017. https://doi.org/10.1155/2017/8345704

[136] Dragomiretskiy K, Zosso D. Variational mode decomposition. *IEEE Transactions on Signal Processing* 2014;62:531–44. https://doi.org/10.1109/TSP.2013.2288675

[137] Wang Y, Markert R, Xiang J, Zheng W. Research on variational mode decomposition and its application in detecting rub-impact fault of the rotor system. *Mechanical Systems and Signal Processing* 2015;60:243–51. https://doi.org/10.1016/j.ymssp.2015.02.020

[138] Zhang S, Zhao H, Xu J, Deng W. A novel fault diagnosis method based on improved adaptive variational mode decomposition, energy entropy, and probabilistic neural network. *Transactions of the Canadian Society for Mechanical Engineering* 2020;44:121–32. https://doi.org/10.1139/tcsme-2018-0195

[139] Apostolidis GK, Hadjileontiadis LJ. Swarm decomposition: A novel signal analysis using swarm intelligence. *Signal Processing* 2017;132:40–50. https://doi.org/10.1016/j.sigpro.2016.09.004

[140] Miao Y, Zhao M, Makis V, Lin J. Optimal swarm decomposition with whale optimization algorithm for weak feature extraction from multicomponent modulation signal. *Mechanical Systems and Signal Processing* 2019;122:673–91. https://doi.org/10.1016/j.ymssp.2018.12.034

[141] Li G, Shi J. On comparing three artificial neural networks for wind speed forecasting. *Applied Energy* 2010;87:2313–20. https://doi.org/10.1016/j.apenergy.2009.12.013

[142] Liu H, Tian HQ, Liang XF, Li YF. Wind speed forecasting approach using secondary decomposition algorithm and Elman neural networks. *Applied Energy* 2015;157:183–94. https://doi.org/10.1016/j.apenergy.2015.08.014

[143] Abdoos AA. A new intelligent method based on combination of VMD and ELM for short term wind power forecasting. *Neurocomputing* 2016;203:111–20. https://doi.org/10.1016/j.neucom.2016.03.054

[144] Liu X, Huang H, Xiang J. A personalized diagnosis method to detect faults in gears using numerical simulation and extreme learning machine. *Knowledge-Based Systems* 2020;195:105653. https://doi.org/10.1016/j.knosys.2020.105653

[145] Kang W, Chen W. Optimizing online sequential extreme learning machine parameters and application to transformer fault diagnosis. *2015 4th International Conference on Mechatronics, Materials, Chemistry and Computer Engineering* 2015:892–7. https://doi.org/10.2991/icmmcce-15.2015.177

[146] Haider Shah S, Iqbal K, Riaz A. Constrained optimization-based extreme learning machines with bagging for freezing of gait detection. *Big Data and Cognitive Computing* 2018;2:31. https://doi.org/10.3390/bdcc2040031

[147] Pang S, Yang X, Zhang X. Aero engine component fault diagnosis using multi-hidden-layer extreme learning machine with optimized structure. *International Journal of Aerospace Engineering* 2016;2016. https://doi.org/10.1155/2016/1329561

[148] Matilla-García M, Ruiz Marín M. A non-parametric independence test using permutation entropy. *Journal of Econometrics* 2008;144:139–55. https://doi.org/10.1016/j.jeconom.2007.12.005

[149] Zhang X, Liang Y, Zhou J. A novel bearing fault diagnosis model integrated permutation entropy, ensemble empirical mode decomposition and optimized SVM. *Measurement* 2015;69:164–79. https://doi.org/10.1016/j.measurement.2015.03.017

[150] Li Y, Xu M, Wei Y, Huang W. A new rolling bearing fault diagnosis method based on multiscale permutation entropy and improved support vector machine based binary tree. *Measurement* 2016;77:80–94. https://doi.org/10.1016/j.measurement.2015.08.034

[151] Li B, Li Y, Rong X. The extreme learning machine learning algorithm with tunable activation function. *Neural Computing and Applications* 2013;22:531–9. https://doi.org/10.1007/s00521-012-0858-9

[152] Zhang X, Yang Z, Cao F, Cao J, Wang M, Cai N. Conditioning optimization of extreme learning machine by multitask beetle antennae swarm algorithm. *Memetic Computing* 2020;12:151–64. https://doi.org/10.1007/s12293-020-00301-w

[153] Li S, Chen H, Wang M, Heidari AA, Mirjalili S. Slime mould algorithm: A new method for stochastic optimization. *Future Generation Computer Systems* 2020;111:300–23. https://doi.org/10.1016/j.future.2020.03.055

[154] Chauhan S, Singh M, Aggarwal AK. Design of a two-channel quadrature mirror filter bank through a diversity-driven multi-parent evolutionary algorithm. *Circuits, Systems, and Signal Processing* 2021. https://doi.org/10.1007/s00034-020-01625-1

[155] Chauhan S, Singh M, Aggarwal AK. Cluster head selection in heterogeneous wireless sensor network using a new evolutionary algorithm. *Wireless Personal Communications* 2021;119:585–616. https://doi.org/10.1007/s11277-021-08225-5

[156] Rezaee Jordehi A. Particle swarm optimisation with opposition learning-based strategy: An efficient optimisation algorithm for day-ahead scheduling and reconfiguration in active distribution systems. *Soft Computing* 2020;24:18573–90. https://doi.org/10.1007/s00500-020-05093-2

[157] Kira K, Rendell LA. The feature selection problem: Traditional methods and a new algorithm. *Proceedings of the 10th National Conference on Artificial Intelligence* 1992;2.

[158] Kira K, Rendell LA. *A Practical Approach to Feature Selection*. Morgan Kaufmann Publishers, Inc.; 1992. https://doi.org/10.1016/b978-1-55860-247-2.50037-1

[159] Urbanowicz RJ, Meeker M, La Cava W, Olson RS, Moore JH. Relief-based feature selection: Introduction and review. *Journal of Biomedical Informatics* 2018;85:189–203. https://doi.org/10.1016/j.jbi.2018.07.014

Index

Pages in *italics* refer to figures and pages in **bold** refer to tables.

174

For Product Safety Concerns and Information please contact our EU
representative GPSR@taylorandfrancis.com
Taylor & Francis Verlag GmbH, Kaufingerstraße 24, 80331 München, Germany